W9-AXU-389

*Tuning*
*and Control Loop*
*Performance*

*Instrument*
*Society of*
*America*

*Monograph Series*

Gregory K. McMillan
Monsanto Company, St. Louis, Missouri

# *Tuning and Control Loop Performance*

*Tuning and Control Loop Performance*
*An ISA Monograph*



The Instrument Society of America
67 Alexander Drive
P.O. Box 12277
Research Triangle Park, NC 27709

Library of Congress Cataloging in Publication Data

McMillan, Gregory K., 1946–
Tuning and control loop performance.

(Monograph series / Instrument Society of America;
4)

Bibliography: p.

1. Process control. 2. Feedback control systems.
I. Title. II. Series: Monograph series (Instrument
Society of America) ; 4.
TS156.8.M4 1983 629.8′3 82-48558
ISBN 0-87664-694-1

Design and production by
Publishers Creative Services Inc.

# *Contents*

**Preface**

**Chapter I—Controller Modes and Loop Components** **3**
- 1.1 Controller Modes *3*
  - 1.1.1 Proportional (Gain) Mode *3*
  - 1.1.2 Integral (Reset) Mode *4*
  - 1.1.3 Derivative (Rate) Mode *6*
- 1.2 Loop Components *9*

**Chapter II—Typical Mode Settings and Tuning Methods** **13**
- 2.1 Typical Mode Settings *13*
- 2.2 Typical Tuning Methods *14*
  - 2.2.1 Ultimate Oscillation Method *14*
  - 2.2.2 First Order Plus Dead Time Method *16*
- 2.3 Mechanics of Tuning *18*

**Chapter III—Definition of Performance Criteria** **23**
- 3.1 Accumulated Error *24*
- 3.2 Peak Error *30*

**Chapter IV—Effect of Process Dynamics** **33**
4.1 Methods for Estimation of Dead Time and Time Constant *33*
4.2 Accuracy of Dead Time and Time Constant Approximation *41*
4.3 Self-Regulating Processes *42*
4.4 Integrating Processes *47*
4.5 Runaway Processes *53*
4.6 Examples *61*
4.6.1 Waste Treatment pH Loop (Self-Regulating) *61*
4.6.2 Boiler Feedwater Flow Loop (Self-Regulating) *66*
4.6.3 Boiler Drum Level Loop (Integrating) *69*
4.6.4 Furnace Pressure Loop (Pseudointegrating) *71*
4.6.5 Exothermic Reactor Temperature Loop (Runaway) *76*
4.6.6 Biological Reactor Concentration Loop (Runaway) *79*

**Chapter V—Effect of Controller Dynamics** **83**
5.1 Parallel Controller Algorithm *83*
5.2 Series Controller Algorithm *85*
5.3 Analog Controllers *88*
5.4 Digital Controllers *90*
5.5 Antireset Windup Algorithms *93*

**Chapter VI—Effect of Measurement Dynamics** **101**
6.1 Measurement Time Constant *101*
6.2 Measurement Dead Time *112*
6.3 Measurement Accuracy and Rangeability *114*
6.4 Examples *117*
6.4.1 Waste Treatment pH Loop (Self-Regulating) *117*
6.4.2 Boiler Feedwater Flow Loop (Self-Regulating) *121*
6.4.3 Boiler Drum Level Loop (Integrating) *123*
6.4.4 Furnace Pressure Loop (Pseudointegrating) *123*

6.4.5 Exothermic Reactor Temperature Loop (Runaway) *126*
6.4.6 Biological Reactor Concentration Loop (Runaway) *129*

**Chapter VII—Effect of Valve Dynamics** **133**
7.1 Valve Time Constant *133*
7.2 Valve Dead Time *141*
7.3 Valve Stroke Accuracy and Rangeability *142*
7.3.1 Laminar Flow Valve *147*
7.3.2 Digital Valve *147*
7.3.3 Pulse Interval Control *149*
7.4 Furnace Pressure Loop Example *154*

**Chapter VIII—Effect of Disturbance Dynamics** **157**
8.1 Disturbance Time Constant *157*
8.2 Disturbance Time Interval (Noise) *159*
8.3 Furnace Pressure Loop Example *161*

**Chapter IX—Effect of Nonlinearities** **165**
9.1 Variable Gain *165*
9.2 Variable Time Constant and Dead Time *171*
9.3 Inverse Response *172*
9.4 Waste Treatment pH Loop Example *174*

**Chapter X—Effect of Interaction** **179**

**Chapter XI—Effect of Advanced Control Algorithms** **183**
11.1 Cascade Control *183*
11.2 Feedforward Control *195*
11.3 Signal Characterization *198*
11.4 Mode Characterization *199*
11.5 Direct Synthesis Controller *202*
11.6 Dead Time Compensation *204*
11.7 Self-Tuning Controllers and Regulators *207*
11.8 Model Predictive Controller *210*

**Chapter XII—Summary** **213**

**References** **216**

Appendix A—Equation for the Accumulated Error 221

Appendix B—Equation for the Peak Error 223

Appendix C—Equations for Mass and Energy Balance 227

Appendix D—Equation for Direct Synthesis Controller 233

Appendix E—Equations for Liquid Composition Control Dynamics 237

Appendix F—Equations for Equivalent Noninteractive Time Constants 241

Appendix G—Equations for Gas Pressure Control Time Constants 243

Appendix H—FORTRAN Listing of Subroutine for Controller Tuning 245

Appendix I—ACSL Program for Cascade Control 249

Appendix J—FORTRAN Listing of Subroutine To Estimate the Effective Dead Time and Time Constant for Equal Noninteractive Time Constants 257

Index 259

# *Figures*

| | | |
|---|---|---|
| 1.1 | Definition of Controller Modes | *4* |
| 1.2 | Effect of Proportional Mode on Response | *5* |
| 1.3 | Effect of Integral Mode on Response | *7* |
| 1.4 | Effect of Derivative Mode on Response | *8* |
| 1.5 | Example of Process Gain Nonlinearity | *10* |
| 1.6 | Block Diagram of Loop Components | *11* |
| 2.1 | CRT Display for Reporting Time Smaller than Half the Loop Period | *19* |
| 2.2 | CRT Display for Reporting Time Larger than Half the Loop Period | *20* |
| 3.1 | Accumulated (Integrated) Error | *25* |
| 3.2 | Block Diagram of Loop Components | *27* |
| 3.3 | Estimation of Performance Criteria | *32* |
| 4.1 | Graphical Method to Estimate Dead Time and Time Constant | *35* |
| 4.2 | Effective Dead Time Factor for Small and Large Time Constants | *38* |
| 4.3. | Effective Dead Time Factor for Several Large Equal Time Constants | *39* |
| 4.4 | Three-Dimensional Accumulated Error for a Self-Regulating Process | *47* |
| 4.5 | Three-Dimensional Peak Error for a Self-Regulating Process | *48* |

4.6 Pseudointegrator Gain Approximation *51*
4.7 Three-Dimensional Accumulated Error for an Integrating Process *53*
4.8 Three-Dimensional Peak Error for an Integrating Process *54*
4.9 Effect of Dead Time on Proportional Band Window *56*
4.10 Effect of Time Constant on Proportional Band Window *57*
4.11 Three-Dimensional Accumulated Error for a Runaway Process *59*
4.12 Three-Dimensional Peak Error for a Runaway Process *60*
4.13 Waste Treatment pH Loop (Self-Regulating) *62*
4.14 Boiler Feedwater Flow Loop (Self-Regulating) *66*
4.15 Boiler Drum Level Loop (Integrating) *70*
4.16 Furnace Pressure Loop (Pseudointegrating) *72*
4.17 Reactor Temperature Loop (Runaway) *76*
4.18 Biological Reactor Concentration Loop (Runaway) *79*
5.1 Parallel Controller Algorithm *84*
5.2 Series Controller Algorithm *86*
5.3 Analog Controller Open Loop Test Results *89*
5.4 Location of Proportional Band for Controller Without Antireset Windup *96–97*
6.1 Closed Loop Response for Large Measurement Time Constant ($TC_m = 100*TD$ for a Self-Regulating Process) *103*
6.2 Ratio of Maximum to Minimum Proportional Band for a Runaway Process *107*
6.3 Effect of Measurement Time Constants on the Flow Drop and Oscillations of Surge *108*
6.4 Waste Treatment pH Loop *118*
6.5 Boiler Feedwater Flow Loop *121*
6.6 Boiler Drum Level Loop *123*
6.7 Furnace Pressure Loop *124*
6.8 Exothermic Reactor Temperature Loop *127*
6.9 Biological Reactor Concentration Loop *130*
7.1 Rate-Limited Open Loop Response *134*

7.2 Closed Loop Response for a Large Valve Time Constant ($TC_v$ = 100*$TD$ for a Self-Regulating Process) *138*
7.3 Surge Control Valve Accessories *140*
7.4 Nonlinear Torque Requirement of an Eccentric Disk Valve *143*
7.5 Effect of Control Valve Hysteresis on Loop Oscillation *144*
7.6 Measured Flow Characteristic of Laminar Flow Valve *148*
7.7 Comparison of Pulse Width and Pulse Interval Modulation *150*
7.8 Flow Characteristic of Pulse Interval Control *152*
7.9 Furnace Pressure Loop *153*
8.1 Furnace Pressure Loop *162*
9.1 Installed Valve Gain of an Equal Percentage Inherent Characteristic *166*
9.2 Installed Valve Gain of a Conventional Butterfly Valve *167*
9.3 Process Gain for a Heat Exchanger *169*
9.4 Conductivity Measurement Gain for Sulfuric Acid Concentration *170*
9.5 Inverse Response of a Self-Regulating Process *172*
9.6 Waste Treatment pH Loop *174*
10.1 Half-Decoupling of Furnace Pressure Control *180*
10.2 Pulverized Coal-Fired Boiler Furnace Pressure Response *181*
11.1 Block Diagram of Cascade Control Loop *184*
11.2 Peak Error Ratio for Self-Regulating Inner and Outer Processes *187*
11.3 Peak Error Ratio for Integrating Inner Processes *188*
11.4 Peak Error Ratio for Runaway Inner Processes *189*
11.5 Peak Error Ratio for Integrating Outer Processes *190*
11.6 Peak Error Ratio for Runaway Outer Processes *191*
11.7 Block Diagram of Feedforward Control Loop *196*
11.8a Full Titration Curve *200*
11.8b Blown-up Titration Curve *201*
11.9 Smith Predictor Block Diagram *205*

11.10 Comparison of Smith Predictor Response with Conventional PID Controller (Self-Regulating Process) *206*
11.11 Comparison of Smith Predictor Response with Conventional PID Controller (Integrating Process) *207*
11.12 Comparison of Smith Predictor Response with Conventional PID Controller (Runaway Process) *208*
11.13 Smith Predictor for Flue Gas CO Control *209*
12.1 Block Diagram of Loop Components *214*
F1 Two Interactive Time Constants *242*

# *Preface*

The purpose of a control loop is to reduce the error between the measurement and the set point. The ability of the loop to reduce this error depends on the controller mode settings. If the engineer can estimate the best possible controller mode settings, the maximum potential of the loop has a better chance of being achieved. Also, the control error can be estimated from these mode settings so that the justification for the investment in the loop design and construction can be judged on the basis of loop performance.

Most of the published work on control loop analysis utilizes such techniques as Laplace transforms, *Z* transforms, and state space analysis, which are unfamiliar to the practicing design and maintenance engineer. The engineer may become involved in the complexity of the techniques and lose a feel for the practical aspects of the problem and sight of the real objectives. Simple algebraic equations are needed that illustrate basic rules of thumb. J. G. Ziegler and N. B. Nichols originated, and F. G. Shinskey popularized, many techniques that have proved useful to the design and maintenance engineer for tuning and loop performance, such as the ultimate oscillation and reaction curve methods for controller tuning, the dead time approximation method for characterizing higher order dynamics, and the integrated error approximation for estimating loop performance. These techniques are extended and new simple equations are presented. The tuning methods documented by P. W. Murrill also are summarized. The algebraic equations and the rules of thumb that they illustrate should be simple enough that tuning and performance estimation can become part of the normal design and troubleshooting activities of engineers.

*Gregory K. McMillan*

# *Acknowledgments*

The author wishes to express his gratitude to Vernon L. Trevathan for introducing the basic concepts of loop performance, to Henry H. Chien for the concept of the pseudointegrator, to Robert E. Otto for the concept of the proportional band window for the runaway process, to Terry L. Tolliver for his help in running ACSL simulation programs, to Norman C. Pereira for his help in gathering literature, to Jill A. Westermayer for her help in generating the TELLAGRAF® plots, and to Linda L. Burks for the word processing.

# *Tuning and Control Loop Performance*

CHAPTER 1

# Controller Modes and Loop Components

## 1.1 Controller Modes

Standard controllers have up to three "modes or terms in their control algorithm. (The word *mode* is used here to classify the output response of the controller to a set point and measurement and does not refer to whether the controller is in manual or automatic.) Figure 1.1 shows the mathematical definition and the two commonly used names for each mode.

### 1.1.1 PROPORTIONAL (GAIN) MODE

Nearly all controllers have the proportional (gain) mode. This mode changes the controller output by an amount proportional to the change in error. The proportional band is the percent change in error necessary to cause a full-scale change in controller output. Proportional band is the inverse of controller gain multiplied by 100. Most of the analog controllers use proportional band whereas the majority of the new digital controllers use gain. Note that the proportional band setting also affects the integral and derivative modes. Figure 1.2 shows that as the proportional band is decreased (as the gain is increased), the offset (sustained error) decreases but the response becomes more oscillatory. If the proportional band is decreased until the oscillations are equal in amplitude and no other modes are used, the period of these oscillations is the natural period of the loop. This natural period is known as the ultimate period and depends on the dynamics of the process and the instrument components of the loop. If the oscillations are growing in amplitude, the measured period is shorter than the ultimate period. If the oscillations are decaying in amplitude, the measured period is longer than the ultimate period.

### 1.1.2 INTEGRAL (RESET) MODE

Most controllers also have an integral (reset) mode. This mode changes the controller output by an amount proportional

Proportional (gain) $\frac{100}{PB} * E$

Integral (reset) $\frac{100}{PB} * \frac{1}{T_i} * \int E dt$

Derivative (rate) $\frac{100}{PB} * T_d * \frac{dE}{dt}$

$PB$ = proportional band (%)

$E$ = error

$T_i$ = integral time (minutes/repeat)

$T_d$ = derivative time (minutes)

FIGURE 1.1
Definition of controller modes.

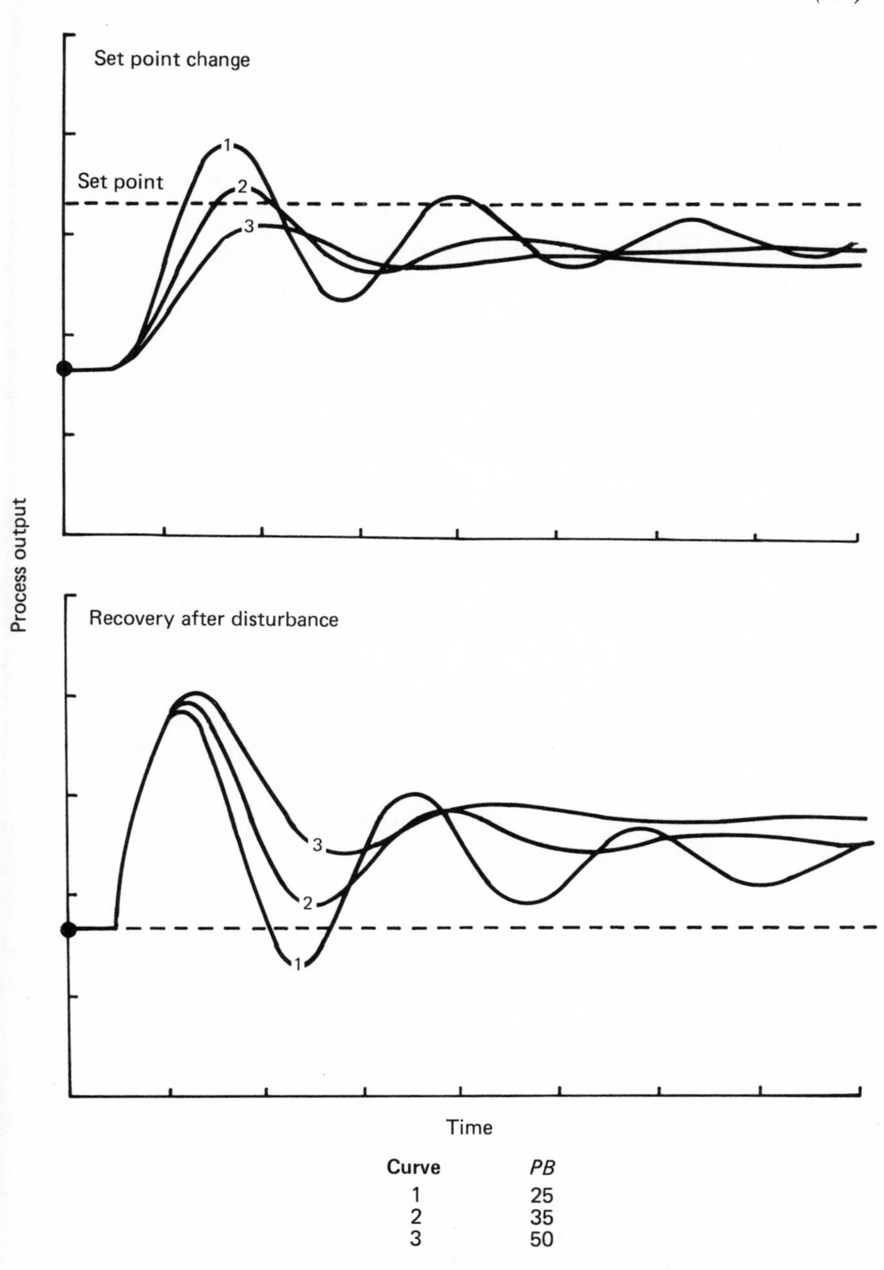


FIGURE 1.2
Effect of proportional mode on response.

to the integral of the error. The integral time is the time required for the integral (reset) mode contribution to equal (repeat) the proportional mode contribution for a constant error. The integral mode action lags the proportional mode action by this integral time. The use of the integral mode increases the permissible proportional band but eliminates offset. Most controllers use the inverse of the integral time so that the mode setting units are repeats per minute. Figure 1.3 shows that as the integral time is decreased (reset action in repeats per minute increased), the offset is eliminated faster but the response becomes more oscillatory. If the integral time is decreased too much, the oscillations develop into a reset cycle whose period is much longer than the ultimate period.

### 1.1.3 DERIVATIVE (RATE) MODE

The derivative (rate) mode is used in only a few loops because the minimum setting available is too large for many loops, derivative action amplifies noise, and the tuning is more complicated as a result of interaction with the other modes in most industrial controllers. This mode changes the controller output by an amount proportional to the derivative of the error. It gives an anticipatory type of response that is useful for slow processes. The derivative time is the time required for the proportional mode contribution to equal the derivative (rate) mode contribution for a ramp error. The derivative action leads the proportional action by this derivative time. If there is no noise, the use of derivative action decreases the permissible proportional band. Figure 1.4 shows that as the derivative time is increased, the overshoot for set point changes and the peak error for load disturbances is reduced but the response becomes more oscillatory. If the response turns back as the measurement approaches set point before if crosses set point, the derivative time is longer than normal. If the derivative time is increased too much, the oscillations develop into a rate cycle whose period is shorter than the ultimate period.

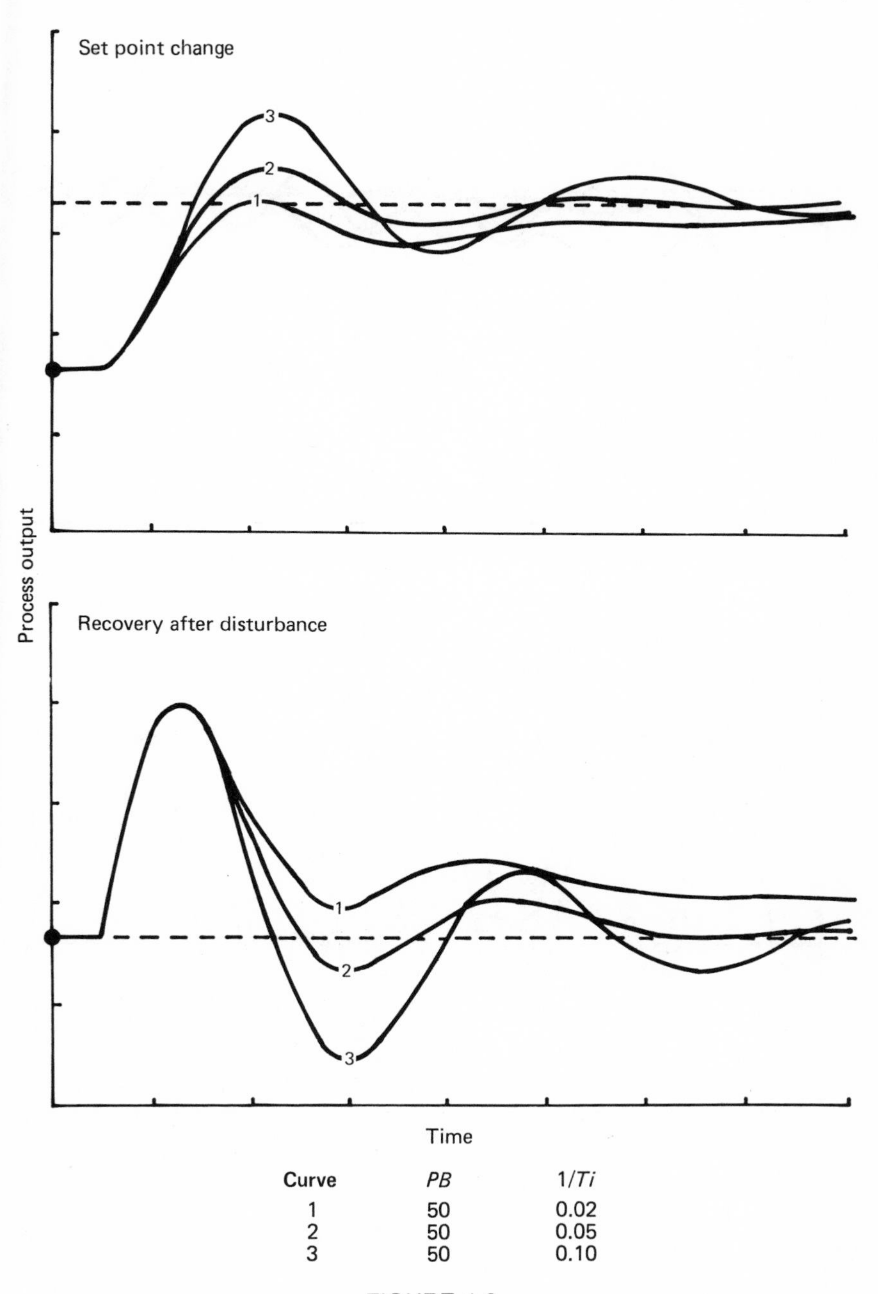


| Curve | *PB* | $1/Ti$ |
|---|---|---|
| 1 | 50 | 0.02 |
| 2 | 50 | 0.05 |
| 3 | 50 | 0.10 |

FIGURE 1.3
Effect of integral mode on response.

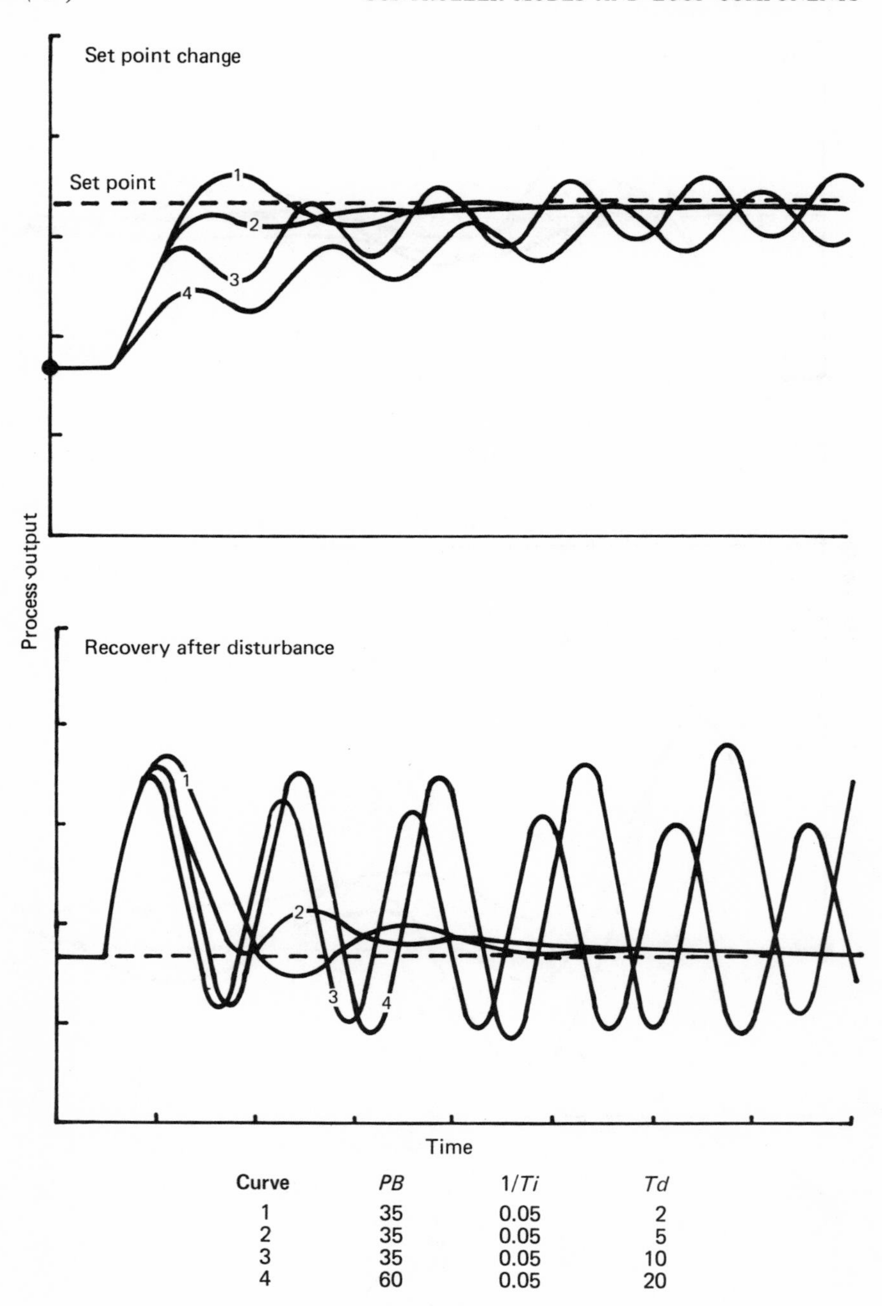


| Curve | *PB* | 1/*Ti* | *Td* |
|---|---|---|---|
| 1 | 35 | 0.05 | 2 |
| 2 | 35 | 0.05 | 5 |
| 3 | 35 | 0.05 | 10 |
| 4 | 60 | 0.05 | 20 |

FIGURE 1.4
Effect of derivative mode on response.

## 1.2
## *Loop Components*

The major components of a loop are the controller, control valve, process, and measurement. Each component has a steady-state gain and one or more dynamic terms to describe its response. The definition of these terms is presented to avoid confusion (their graphic approximation is illustrated in Chapter 4).

*Dead time (time delay):* time required for the output to start to change after a change in input.

*Integrator gain:* slope of the ramp change in output for a step change in input.

*Negative feedback time constant:* time required for the output to reach 63 percent of the input multiplied by its gain after the output starts to change for a step change in input. The output approaches a new steady state with exponentially decreasing slope.

*Positive feedback time constant:* time required for the output to reach 172 percent of the input multiplied by its gain after the output starts to change for a step change in input. The output approaches infinity or a physical limit with exponentially increasing slope.

*Steady-state gain:* final change in output divided by the change in input after all the transients have died out. It is the slope of a steady-state plot of output versus input. If the plot is a straight line, the gain is linear (slope is constant). If the plot is a curve, the gain is nonlinear (slope varies with operating point). Figure 1.5 shows how the pH process gain varies with the pH measurement signal. [The slope of the titration curve varies with pH (McMillan, 1979).]

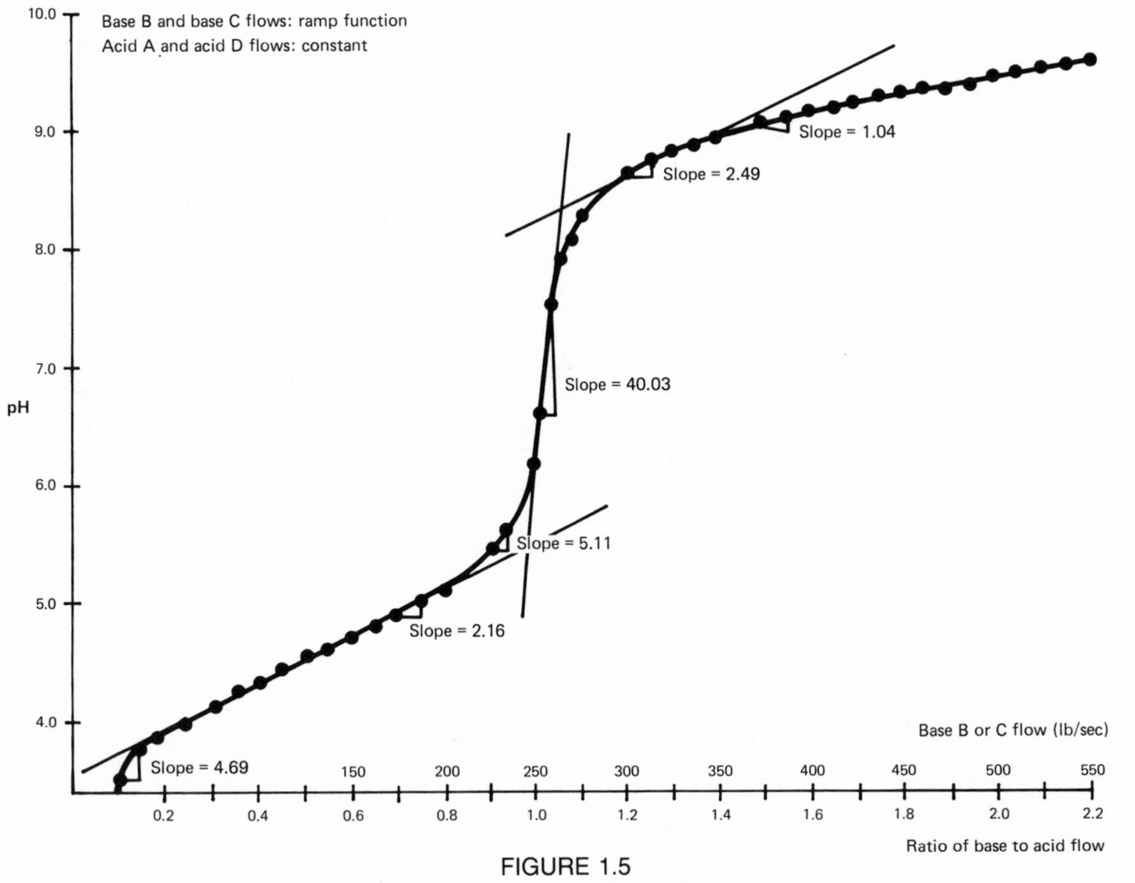


FIGURE 1.5
Example of process gain nonlinearity.

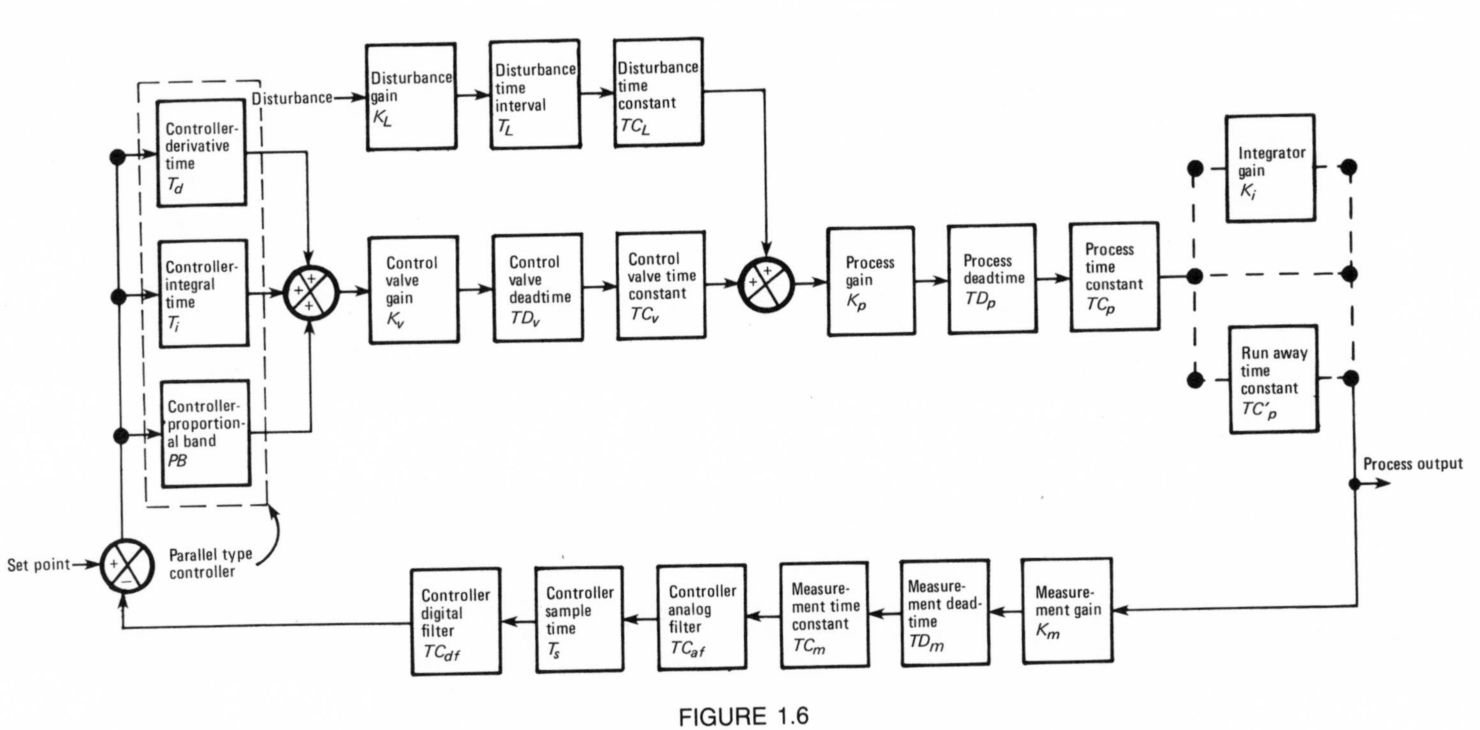


FIGURE 1.6
Block diagram of loop components.

Figure 1.6 is a block diagram of the major components of a loop with the steady-state gain and dynamic terms identified. Most of the literature to date has concentrated on the effect of the gain and dynamic terms of self-regulating (negative feedback) processes. The effect of integrating and runaway (positive feedback) process terms and of instrument terms on tuning and loop performance has not been adequately discussed. A minority of the loops have integrating or runaway terms but most have nonnegligible instrument terms. Instrument dead time and time constant terms affect tuning and loop performance unless they are much smaller than the process dead time (this rule is illustrated in subsequent chapters). The process dead time is the result of transportation delays and multiple time constants in series. Most loops do control around a single piece of process equipment so that the process transportation delays and secondary time constants are small (the most notable exception is distillation columns). Since the process dead time is typically small, the instrument dead time and time constant terms are important. The following is a partial list of control loops whose instrument dynamics affect tuning and loop performance.

1. Flow control
2. Level control
3. Liquid pressure control
4. Gas pressure control of pipeline
5. Gas pressure control of single volume
6. Concentration control of static mixer
7. Concentration control of well-mixed tank*
8. Concentration control using chromatograph
9. Temperature control by exchanger bypass
10. Temperature control of furnace
11. Temperature control of reactor*
12. Compressor surge control

* If the time to turn over the contents of the tank is less than 1 minute (see Appendix E).

CHAPTER II

# Typical Mode Settings and Tuning Methods

## 2.1 Typical Mode Settings

Most loops are actually tuned in practice by setting the modes to some initial typical values. The settings are then trimmed by trial and error on the basis of the observed closed loop response. The desired response frequently depends on what the operator or technician is used to and is typically slower and less oscillatory than the response for minimum error.

Table 2.1 shows typical controller mode settings for various types of loops. These settings assume that the instrument engineer has selected the control valve and measurement spans so that a full-scale stroke gives a full-scale change in measurement signal for self-regulating processes. Those loops whose controllers frequently have to be detuned (proportional band increased and derivative time decreased) because of process noise are flagged with an asterisk.

Gas pressure and level loops typically have an integrating response. The integral mode should be avoided unless the derivative mode can also be used. Neither of these modes is needed if the proportional band is less than 10 percent. Furnace and dryer gas pressure loops are frequently noisy and have high measurement gains since the measurement span may be a fraction of an inch of water column. Such gas pressure loops require large proportional bands. Liquid level loops on distillation columns and boiler drum levels (shrink and swell) may be noisy and have an inverse response where the initial response is opposite to the final response. Liquid level loops with bubblers may be noisy. Such liquid level loops require a large proportional band.

## 2.2 *Typical Tuning Methods*

The tuning methods described in the literature to date are for loops with self-regulating processes. The three major classes of methods are the ultimate oscillation, first order plus dead time, and second order plus dead time. The last class will not be discussed because the better estimate of derivative mode setting is usually not worth the extra effort.

### 2.2.1 ULTIMATE OSCILLATION METHOD

The ultimate oscillation method is carried out with the controller in service. No estimates of loop gain, dead time, or time

TABLE 2.1
Typical Controller Mode Settings

| *Loop Type* | *PB Setting, %* | *Ti Setting, r/m* | *Td Setting, m* |
|---|---|---|---|
| Flow* | 100–500 | 10–50 | none |
| Liquid pressure* | 100–500 | 10–50 | none |
| Gas pressure* | 1–50 | 2–10 | 0.02–0.1 |
| Level* | 1–50 | 4–20 | 0.01–0.05 |
| Temperature | 10–100 | 0.02–1 | 0.5–20 |
| Chromatograph | 200–800 | 0.01–0.1 | none |

%—percent (proportional band)
r/m—repeats per minute (reset)
m—minutes (rate)

constant are required. The method also works well for dead time dominant loops whereas the other classes do not. In this respect, it is nearly foolproof for self-regulating processes. However it requires that the loop response develop undamped oscillations that may be undesirable from an operational or safety viewpoint. It may be difficult to prevent these oscillations from growing or from reaching some physical limit. The ultimate oscillation method developed by J. G. Ziegler and N. B. Nichols (Ziegler & Nichols, 1942) is summarized as follows:

1. Adjust the proportional band dial to its maximum and the reset (repeats per minute) and derivative dials to their minimums. If the reset dial is in minutes per repeat, set it to its maximum (no reset action).
2. Adjust the manual output of the controller to give a measurement as close to midscale as possible.
3. Switch the controller to automatic.
4. Reduce the proportional band setting until the observed oscillations neither grow nor diminish in amplitude. If the oscillations saturate at either extreme, start over at step 2 to stabilize the response. If there are not enough disturbances to start the oscillations, jog the set point. The oscillations have to be only approximately equal in amplitude.
5. Note the proportional band and measure the period of these oscillations. If the recorder speed is too slow, use a stopwatch to time the interval between the first and third pass of the measurement by the set point on the controller scale. If cathode-ray tube (CRT) trend plots are used for fast loops, check the reporting time. The reporting time of data highways may cause low-frequency aliasing (Heider, 1982).
6. Use the ultimate proportional band ($PB_u$) and the ultimate period ($T_u$) from step 5 to estimate the controller mode settings.

Proportional only:

$$PB = 1.8*PB_u$$

Proportional plus integral (PI):

$$PB = 2.22*PB_u \text{ percent}$$
$$T_i = 0.83*T_u \text{ minute per repeat}$$

Proportional plus integral plus derivative (PID):

$$PB = 1.67*PB_u \text{ percent}$$
$$T_i = 0.50*T_u \text{ minute per repeat}$$
$$T_d = 0.125*T_u \text{ minute}$$

The method presented in subsequent chapters is an extension of this method. Equations are presented in Chapter 4 to estimate the ultimate period and the ultimate proportional band for self-regulating, integrating, and runaway processes. Chapters 5, 6, and 7 describe how to modify the terms in these equations to account for the effect of instrument dynamics.

### 2.2.2 FIRST ORDER PLUS DEAD TIME METHOD

The first order plus dead time method is suitable for processes with dead time to time constant ratios of less than one. The reaction curve method is a first order plus dead time method. The graphic technique and the parameters for these methods documented by P. W. Murrill et al. (Murrill et al., 1967) are summarized as follows:

1. Adjust the controller manual output to give a midscale measurement.

2. Move the manual output of the controller by 10 percent rapidly and simultaneously mark the start time on the recorder chart. If the change in measurement is too large or too small, the size of the manual output change will have to be changed accordingly.

3. After the measurement has reached a new steady state, draw a tangent to the inflection point of the open loop response curve. The time from the start mark to the intersection of the tangent with the original measurement is the loop dead time (*TD*). The time from the end of the dead time to the intersection of the tangent with the final measurement is the time constant (*TC*). If the recorder chart speed is too slow, the dead time cannot be graphically determined. A stopwatch can be used to time the interval between the movement of the manual output and the measurement to reach 5 percent of its final value. This

dead time is then subtracted from the time constant that was determined graphically. If CRT trend plots are used for fast loops, check the reporting time. The reporting time of the data highway may be longer than the loop dead time (Heider, 1982).

4. Use the loop dead time and time constant from step 3 to estimate the controller mode settings (see table 2.2).

$$PB = A^*(TD/TC)^{-B}$$

$$PB = A^{-1} * (TD/TC)^{+B} * 100$$

$$T_i = C^*(TD/TC)^{D*}TC$$

$$T_d = E^*(TD/TC)^{F*}TC$$

The ZN settings yield a slightly lower peak error, the IAE settings yield a minimum integrated absolute error, and the ISE settings yield a minimum integrated squared error.

The above proportional band is for an open loop gain of unity. If the open loop gain is not unity, it must be estimated and the above proportional band must be multiplied by it. The ultimate oscillations method did not require knowledge of the open loop gain. The open loop gain is the product of the steady state valve, process, and measurement gains (see equations 3.5 and 3.6 in chapter 3). The open loop gain can be estimated from step 2 of the first order plus dead time method by noting the change in measurement and dividing this change in percent by the percent change in manual controller output. Note that the

TABLE 2.2
First Order Plus Dead Time Method Parameters

| *Method* | *Mode* | *A* | *B* | *C* | *D* | *E* | *F* |
|---|---|---|---|---|---|---|---|
| ZN | P | 1.000 | 1.000 | | | | |
| IAE | P | 0.902 | 0.985 | | | | |
| ISE | P | 1.411 | 0.917 | | | | |
| ZN | PI | 0.900 | 1.000 | 3.333 | 1.000 | | |
| IAE | PI | 0.984 | 0.986 | 1.644 | 0.707 | | |
| ISE | PI | 1.305 | 0.959 | 2.033 | 0.739 | | |
| ZN | PID | 1.200 | 1.000 | 2.000 | 1.000 | 0.500 | 1.000 |
| IAE | PID | 1.435 | 0.921 | 1.139 | 0.749 | 0.482 | 1.137 |
| ISE | PID | 1.495 | 0.945 | 0.917 | 0.771 | 0.560 | 1.006 |

Method key:
ZN—Ziegler–Nichols
IAE—Integrated absolute error
ISE—Integrated squared error

Mode key:
P—Proportional only
PI—Proportional plus integral
PID—Proportional plus integral plus derivative

resulting open loop gain is dimensionless since both the controller measurement input and manual output are expressed in percent. Whether designing your own digital controller algorithm or applying a standard digital or analog industrial controller, it is wise to use percent for controller input and output signals to avoid confusion during the estimation of the open loop gain and controller proportional band.

## 2.3 *Mechanics of Tuning*

The tuning of control room analog controllers is typically accomplished by adjusting small dials or knobs on the side of the controller after it has been partially withdrawn from the case. The adjustment of the dials or knobs in field controllers requires the opening of the weather protection housing. The dials or knobs have continuous or discrete settings. The range of the settings depends on the controller manufacturer and model. The controller may also have switches that multiply the settings by a power of 10. It is difficult to ascertain the mode setting to more than two significant digits owing to the size of the dials or knobs, the lack of intermediate scale graduations, and the inaccuracy of the mode settings.

Digital controllers in distributed control systems typically are tuned by entering digital numbers via a keyboard on a console or a small hand-held tuner. The mode settings are changed in discrete increments but the size of the increment is typically so small that the resolution of mode settings is much greater than that for analog controllers. The mode setting is displayed digitally to three or more significant digits.

The units of the mode settings should be checked and compared with the units of the trend plot display. The proportional mode units are either in percent for proportional band or dimensionless for gain. The integral mode units are either repeats per unit of time or units of time per repeat. The derivative mode units are units of time. The time unit is typically minutes for both the integral and derivative modes.

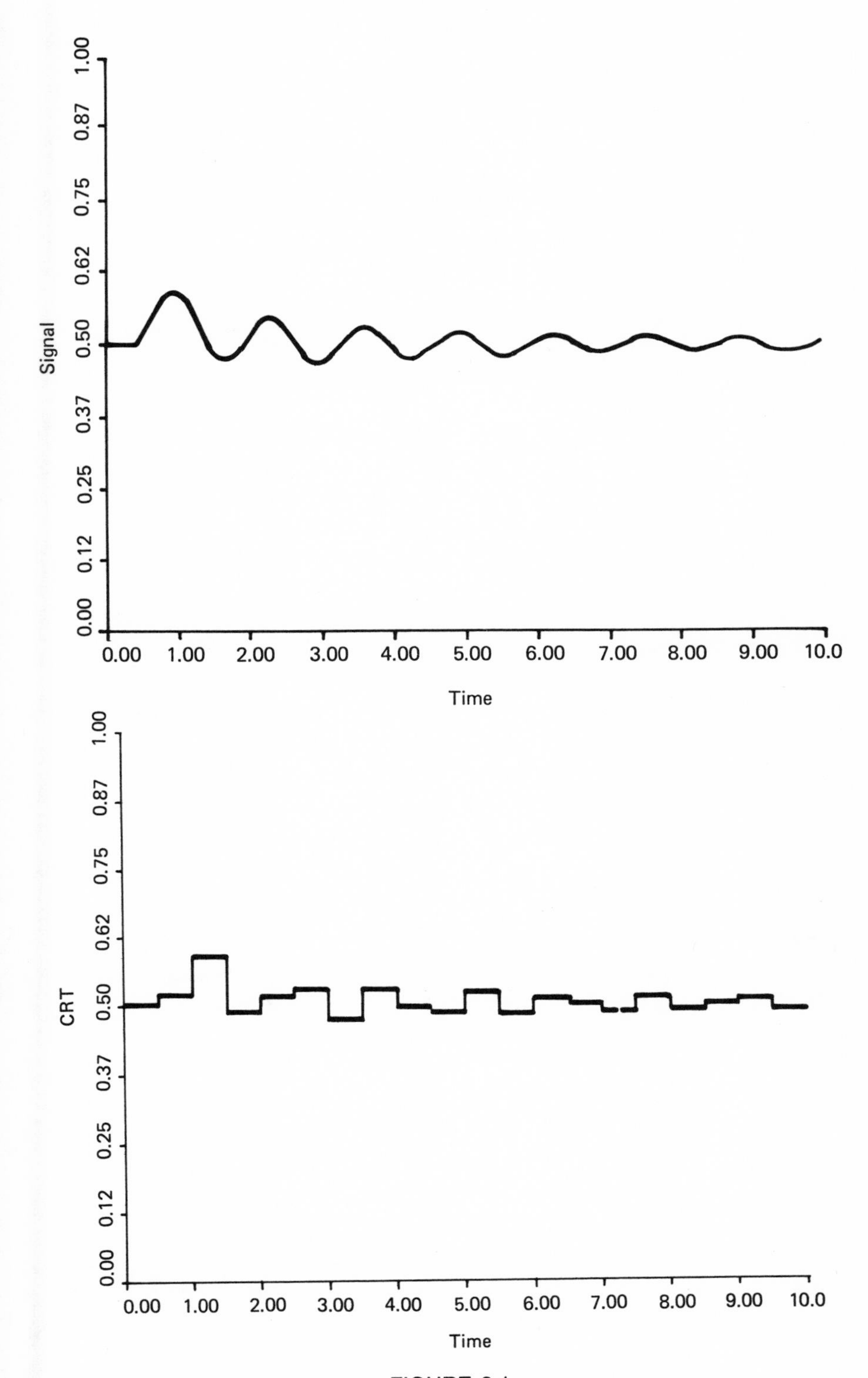


FIGURE 2.1
CRT display for reporting time smaller than half the loop period.

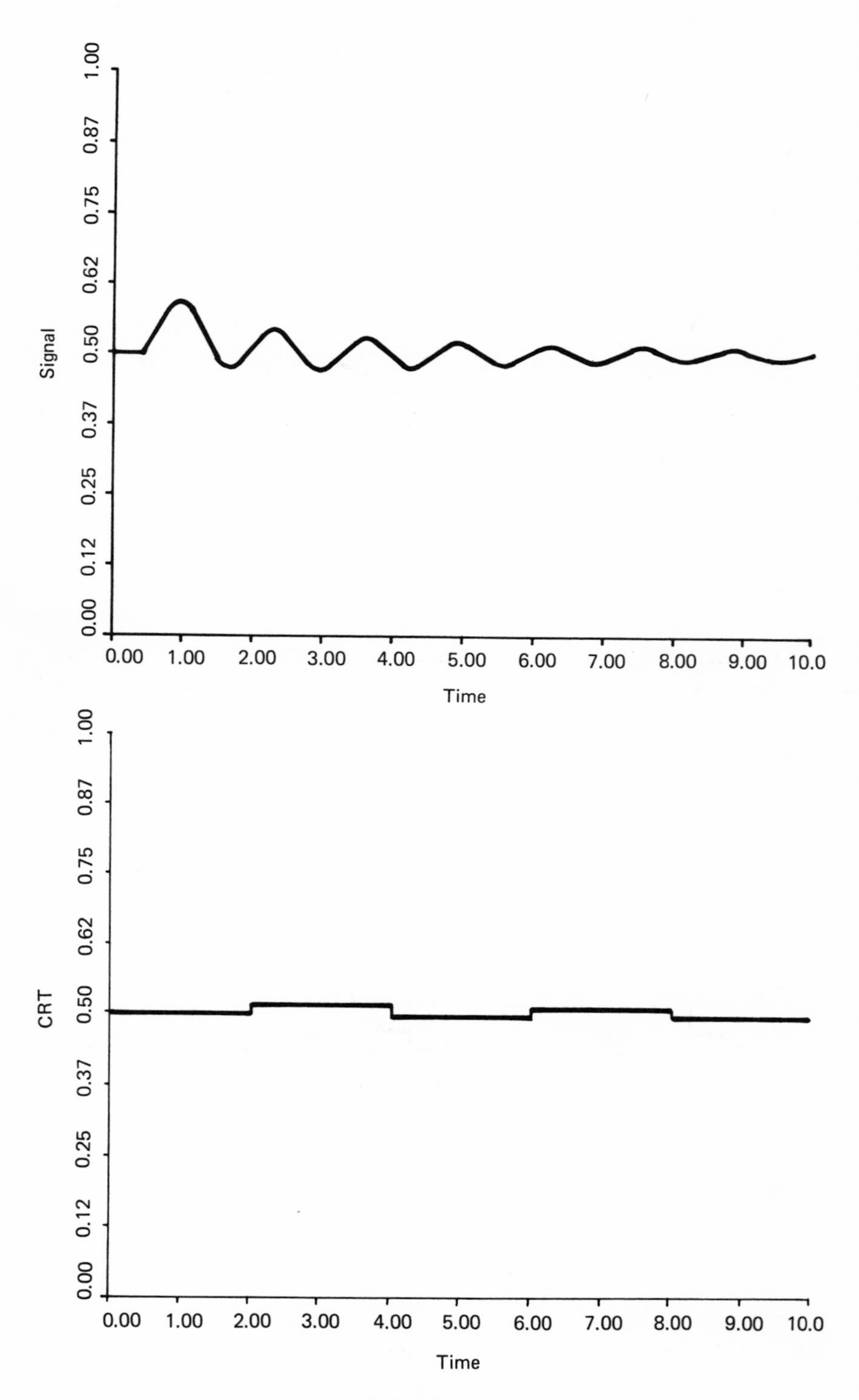


FIGURE 2.2
CRT display for reporting time larger than half the loop period.

The speed of panel strip chart recorders and the reporting time of data highways for CRT displays may be too slow for fast loops. If the performance of the fast loop is important (e.g., reactor feed, compressor surge, or furnace pressure control), a high-speed recorder or data logger should be used to generate a trend plot to show the ultimate period or loop dead time. Oscillographic recorders with thermal or photosensitive paper are reliable and flexible enough to be a useful tool for tuning and troubleshooting critical fast loops. The recently developed Dianachart by the Dianachart Corporation is a good example of a new generation of microprocessor-based, high-speed data loggers that provide plotting and fast event analysis. These data loggers are more powerful, more portable, and less expensive than oscillographic recorders.

Figures 2.1 and 2.2 show the effect of data highway reporting time on the trend plot of a damped control loop oscillation on a CRT display. Notice how the reporting time lengthens the dead time estimate (the step load disturbance occurs at zero time). The pattern of the oscillation also becomes less recognizable and less distinguishable from noise. Figure 2.2 shows the generation of a low-frequency alias (see Chapter 5).

CHAPTER III

# Definition of Performance Criteria

Engineers tend to emphasize qualitative criteria such as loop importance and ease of tuning rather than quantitative criteria such as error size and duration. This qualitative emphasis is due partly to the complexity and diversity of the quantitative criteria and the associated analysis techniques. For example, a level loop that has a non-self-regulating response may be judged easy to control even though the errors may be large in size and duration because these errors are unimportant as long as the tank does not run dry or overflow. A temperature loop with a large time constant may be judged difficult to control even though the errors are small because the slowness of the loop response makes tuning extremely tedious. To judge objectively whether a loop is easy to control, quantitative criteria that are generally applicable and readily understood should be used. All quantitative criteria can be broadly classified as, and simplified to, either an accumulated error or peak error criteria or a combination thereof.

## *3.1 Accumulated Error*

Accumulated error is the totalized deviation of the controlled variable from set point. For a composition control loop (e.g., a distillation column temperature or pH control loop), the accumulated error multiplied by the average product flow provides a measure of the total amount of the product that deviates from the optimum product purity specification. For a flow control loop (e.g., a reactor feed or a furnace fuel loop), the accumulated error provides a measure of the total amount of feed that deviates from the stoichiometric ratio specification that may be important not only for product purity, but also for safety (if the excess feed is flammable, explosive, exothermically reactive, or toxic) or emission control (if the excess feed or associated by-products are environmentally restricted). If the controlled variable is a utility flow (e.g., steam flow or cooling water flow), the accumulated error is representative of energy usage in excess of the set point. The accumulated error is the integrated error where the positive and negative errors are canceled by the volume of the system to provide a net accumulated positive or negative error. The accumulated error in Figure 3.1 is the total positive shaded area minus the total negative shaded area. Integrated absolute error (IAE) is equal to the accumulated error for an overdamped (nonoscillatory) response. Integrated squared error (ISE) can be approximated by a combination of the accumulated error and peak error. The additional value of calculating the IAE and ISE errors is usually not worth the additional effort by the design or maintenance engineer for loop performance analysis. However, a small measured accumulated error does not necessarily mean a well-tuned stable loop. A small measured accumulated error can result from a loop that is marginally stable since the positive and negative errors will cancel for sustained oscillations. The accumulated error can be accurately calculated by use of a relatively simple algebraic equation if the oscillations are decaying in amplitude.

The accumulated (integrated) error for a closed loop can be calculated for a step disturbance if the proportional band, measurement gain, and reset settings of the controller are known

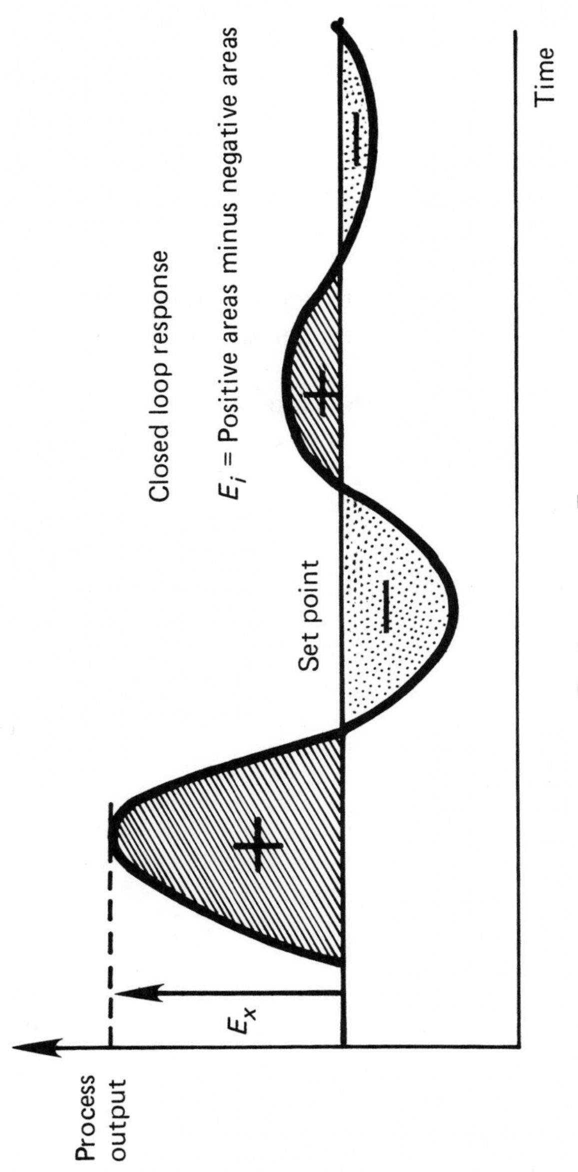


Peak error = $E_x$

Integrated error = $\int E dt$

FIGURE 3.1
Accumulated (integrated) error.

or can be estimated from the steady-state gains and the dynamics of the loop (Shinskey, 1979).

See Appendix A for details on the origin of the following equation.

$$E_i = \frac{PB}{100*K_m}*T_i*\Delta C \tag{3.1}$$

where

$E_i$ = accumulated (integrated) error of the controlled variable ($E_i$ is also the accumulated error of the measured variable if $K_m = 1$)
$PB$ = proportional band of the controller (100 percent/gain)
$T_i$ = reset time of the controller (minutes/repeat)
$K_m$ = measurement of steady-state gain of the transmitter
$\Delta C$ = controller output change required

The controller output change $\Delta C$ required to compensate for the disturbance is equal to the disturbance (process load change $\Delta L$) if both equally affect the controlled variable (the control valve gain $K_v$ equals the load gain $K_l$). See Figure 3.2 for the location of these gain terms in a block diagram of a control loop.

$$\Delta C = \frac{K_l}{K_v}*\Delta L \tag{3.2}$$

$$E_o = K_p*K_l*\Delta L \tag{3.3}$$

Substitute Equations 3.2 and 3.3 into 3.1 and multiply the numerator and denominator by $K_p$:

$$E_i = \left[\frac{PB}{100*K_v*K_p*K_m}\right]*T_i*E_o \tag{3.4}$$

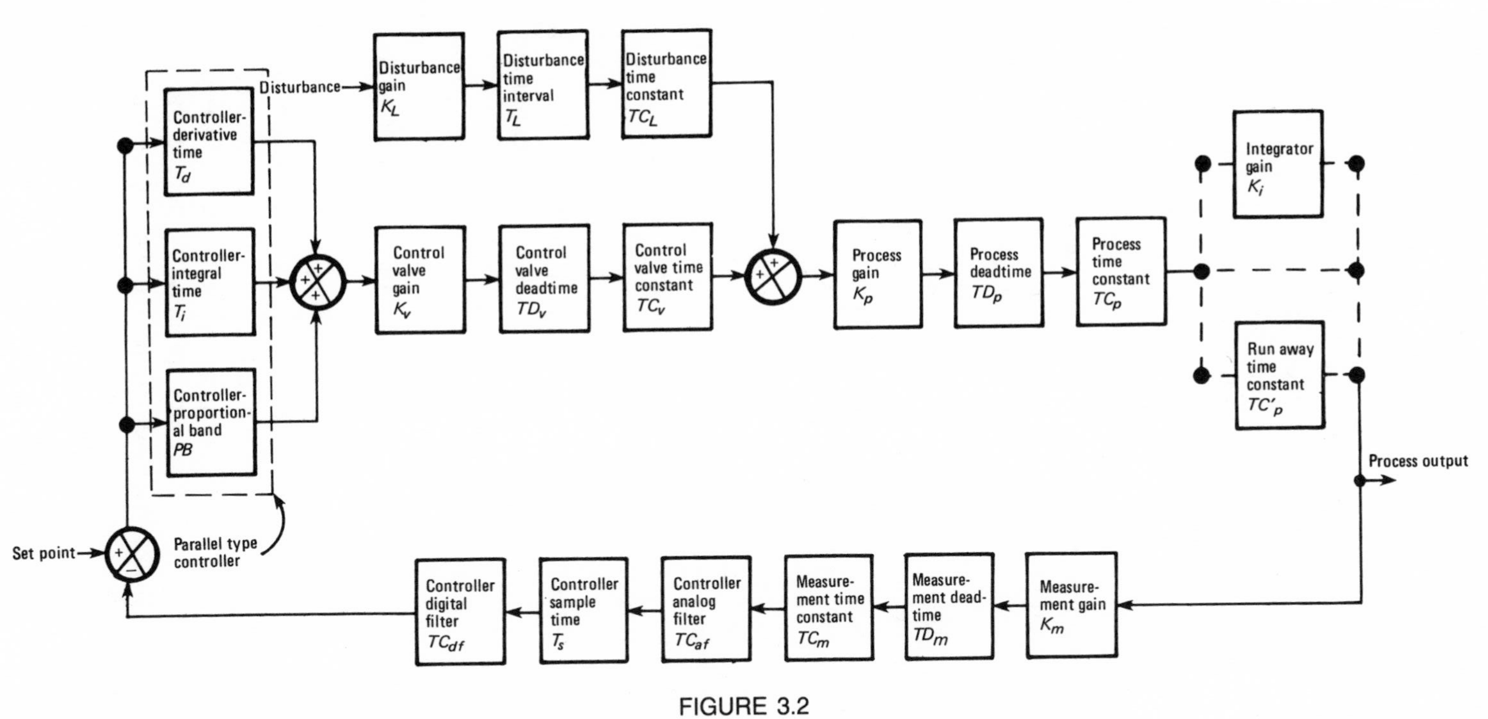


FIGURE 3.2
Block diagram of loop components.

where

$\Delta C$ = controller output change required
$K_v$ = control valve steady-state gain
$K_l$ = process load steady-state gain
$\Delta L$ = process load change (disturbance size)
$K_p$ = process steady-state gain
$E_o$ = open-loop steady-state error

The equation for accumulated error leads to several conclusions important enough to emphasize. First, if the disturbances are nearly zero in magnitude ($E_o = 0.0$), even the most difficult loop will perform excellently. Thus before one can decide whether a difficult loop justifies the additional expense of special equipment, special instruments, or advanced control algorithms, knowledge of the size of the disturbance is necessary. Second, if the controller was tuned with too large a proportional band (too small a gain) or too large a reset time (too slow a reset action), an easy loop will perform as poorly as a more difficult loop that required the higher mode settings. Any special efforts or expense during design to improve loop performance will be wasted if overly conservative controller tuning is used in the field. Third, if the resolution or rangeability of the mode settings of the controller prevents the use of both the best proportional band and reset time settings, an easy loop will again perform as poorly as the more difficult loop. Any capital expended for hardware and design to improve loop performance is wasted where *PB* settings or reset time settings are required that are below the lower limits of the available controller. Fourth, if the process gain is increased, the open loop error, and hence the accumulated error, increases. (The $K_p$ in the denominator is canceled by the $K_p$ in the numerator expression for *PB* since *PB* is proportional to $K_p$.) It is therefore important that the process engineer and control engineer review the effect of equipment design and operating conditions on the process gain at an early stage of the project (preferably before the project appropriation request is made final). However, an increase in proportional band resulting from an increase in valve or measurement gain does not result in an increase in accumulated error of the controlled variable for a given load change since the gain product $K_v$*$K_m$*$K_p$ in the equation for the proportional band (see Chapter 4) cancels out the gains in the denominator of Equation 3.4. Thus it is important for loop performance that the instrument gains be maximized and the process and load gains be minimized.

Many loops are designed by the instrument engineer to provide a full-scale transmitter output change for a full-scale valve stroke that yields the simpler Equation 3.7. Although the overall gain may be unity, there are frequently localized increases in gain resulting from nonlinearities in the control valve gain (e.g., equal percentage trim), process gain (e.g., boiler drum level), and transmitter gain (e.g., orifice flow measurement). These localized increases in gain can cause localized oscillations unless the proportional band is increased (see Chapter 9).

$$\frac{\Delta V}{\Delta C} * \frac{\Delta P}{\Delta V} * \frac{\Delta M}{\Delta P} = 1 \tag{3.5}$$

$$K_v * K_p * K_m = 1 \tag{3.6}$$

Substitute Equation 3.6 into 3.4:

$$E_i = \frac{PB}{100} * T_i * E_o \tag{3.7}$$

where

$\Delta V$ = change in valve output
$\Delta C$ = change in controller output
$\Delta P$ = change in process output
$\Delta M$ = change in measurement output
$K_v$ = steady-state valve gain ($\Delta V/\Delta C$)
$K_p$ = steady-state process gain ($\Delta P/\Delta V$)
$K_m$ = steady-state measurement gain ($\Delta M/\Delta P$)
$E_o$ = open-loop steady-state error
$E_i$ = accumulated (integrated) error

The derivative mode setting is not needed in the equations for the accumulated error because the addition of derivative action decreases the "real" proportional band and increases the "real" reset time by the same factor, which means the accumulated error remains the same (see Chapter 5 for more details).

## 3.2 Peak Error

Peak error is the maximum deviation of the controlled variable from set point. For a reactor temperature or reactor pH loop, the maximum deviation might have to be limited to prevent the start of an undesirable secondary reaction. For a pressure control loop, the maximum deviation might have to be limited to prevent actuation of relief valves or vacuum breakers. For a Claus furnace combustion control loop, the maximum emission level of sulfur dioxide might have to be limited to meet environmental regulations (a Claus furnace produces sulfur from sulfur dioxide waste gas). The peak error is equal to approximately 1.5 times the steady-state offset for a proportional-only controller (Harriot, 1964).

$$E_x = \left[\frac{1.5}{1 + K_o}\right] * E_o \qquad (3.8)$$

where

$E_x$ = peak (maximum) error of the controlled variable

$K_o$ = overall loop gain $\left(K_o = K_v * K_p * K_m * \frac{100}{PB}\right)$

$E_o$ = open-loop steady-state error ($E_o = K_p * K_l * \Delta L$)

Substitute the expression for $K_o$ into Equation 3.8 and multiply the numerator and denominator by $PB$:

$$E_x = \left[\frac{1.5 * PB}{100 * K_v * K_p * K_m + PB}\right] * E_o \qquad (3.9)$$

The addition of reset action usually does not affect the peak error appreciably since the time duration of the peak is small. If the proportional band is small ($PB << K_v*K_p*K_m*100$), then the $PB$ in the denominator of Equation 3.9 can be eliminated, which yields Equation 3.10. Simulation results show that Equation 3.10 is actually more accurate than Equation 3.9 for a controller with rate action (see Appendix B for details on the origin of the following equation).

$$E_x = \left[\frac{K*PB}{100*K_v*K_p*K_m}\right]*E_o \qquad (3.10)$$

Substitute Equation 3.6 into 3.10:

$$E_x = \left[\frac{K*PB}{100}\right]*E_o \qquad (3.11)$$

where

$E_x$ = peak (maximum) error of the controlled variable ($E_x$ is also the peak error of the measured variable if $K_m = 1$)
$K$ = proportionality constant ($K = 1.1$ for quarter's amplitude damping)
$PB$ = proportional band of the controller (100 percent/gain)
$T_i$ = reset time of the controller (minutes/repeat)
$K_v$ = control valve steady-state gain
$K_p$ = process steady-state gain
$K_m$ = measurement steady-state gain
$E_o$ = open-loop steady-state error

Study of the equation for the peak error shows that it is equal to the accumulated error multiplied by $K/T_i$. All of the conclusions discussed for the accumulated error on the effect of disturbance size, controller tuning, proportional band resolution and rangeability, and loop component gains also apply to the peak error. The accuracy of the equation for the peak error is not quite as good as that for the accumulated error since the proportionality constant $K$ changes with the degree of damping (the ratio of dead time to time constant does not affect $K$

appreciably unless this ratio is larger than 0.5, which rarely occurs in chemical processes). If noise (e.g., flow loop) or excessive dead time (e.g., chromatograph loop) necessitates the use of large proportional bands, then Equation 3.8 is more accurate than Equation 3.10 for self-regulating loops. $K$ is 1.1 in Equation 3.10 if the amplitude of the second peak is one-fourth the amplitude of the first peak. This damping of the oscillations is referred to as quarter amplitude. True quarter amplitude damping does not occur except for set point changes or large dead time to time constant ratios. For load changes and small dead time to time constant ratios that are typical for chemical processes, the amplitude of the half cycle after the first peak is much less than that predicted by quarter-amplitude damping, even though the amplitude of the second peak (one full cycle after the first peak) is one fourth of the amplitude of the first peak.

Figure 3.3 summarizes the equations used to estimate the performance criteria of accumulated and peak errors.

$$E_x = \left[\frac{K^*PB}{100^*K_v{}^*K_p{}^*K_m + PB}\right]{}^*E_o$$

$$E_i = \left(\frac{PB}{100^*K_v{}^*K_p{}^*K_m}\right){}^*T_i{}^*E_o$$

$E_x$ = peak (maximum) error of the controlled variable

$E_i$ = accumulated (integrated) error

$PB$ = proportional band of the controller

$K_v$ = control valve steady-state gain

$K_p$ = process steady-state gain

$K_m$ = measurement steady-state gain

$E_o$ = open-loop steady-state error

$T_i$ = reset time of the controller

FIGURE 3.3
Estimation of performance criteria.

CHAPTER IV

# *Effect of Process Dynamics*

## *4.1 Methods for Estimation of Dead Time and Time Constant*

Chemical process dynamics typically consist of a large number of time constants in series. Multiple time constants in series result in an interval of time during which the process does not respond after an input change. This time interval of negligible response is called dead time (also known as time delay). Ziegler and Nichols (Ziegler & Nichols, 1943) developed graphical techniques and equations that characterized the response of several time constants in a series as a combination of a dead time and a single time constant. If the open loop response of the process to a step change (the time from start to finish of the change should be less than 10 percent of the dead time) in either process load $\Delta L$ or controller output $\Delta C$ can be recorded, then the graphical technique can be used. A tangent is drawn to the first inflection point and extended to intersect the time axis. The time between the start of the disturbance and the intersection of the tangent with the time axis is the effective dead time (time delay). The time between the intersection of the tangent with the time axis and the intersection of the tangent with the final value of the response is the effective time constant. If the slope of the response is steep at 60 percent, the loop probably contains multiple equal time constants. For such loops, the effective dead time is graphically estimated the same but the effective time constant is the time between the intersection of the tangent with the time axis and the intersection of the tangent with 60 percent of the open loop error divided by a correction factor. The correction factor is equal to one plus the ratio of the dead time to the time constant before correction. If the equal time constants are known, the FORTRAN subroutine in Appendix J can be used to estimate the effective dead time and the effective time constant.

Figure 4.1 shows the open loop response for the three major types of processes. The self-regulating process (curve 0) is "S" shaped and levels off at a new steady state. It is fairly easy graphically to approximate the dead time and time constant for this type of process by sketching the tangent to the inflection point. The integrating process (curve 1) does not level off at a new steady state but ramps off until a physical limit is reached. The slope of the ramp is the integrator gain. The tangent is not easily constructed because the inflection point cannot be readily identified. Since there is no final value, the time constant is the time interval between the intersection of the tangent with the time axis and with an open error ($E_o$). The runaway process (curve 2) also does not level off at a steady-state value but turns upward and rapidly increases. The negative feedback time constant (*TC*) is identified by the same method used for the integrating process while the positive feedback time constant (*TC*′) is identified by constructing a second tangent. The time interval between the intersection of the time axis and intersection with 172 percent of the open loop error ($E_o$) by this additional tangent is the positive feedback time constant (*TC*′).

If the time constants of the process can be estimated from data on equipment sizes and operating conditions (see Table 4.1), then the Table 4.1 equations or their plots can be used to estimate the effective dead time and time constant.

If the process consists of one large time constant and several smaller time constants, the effective dead time is equal to the

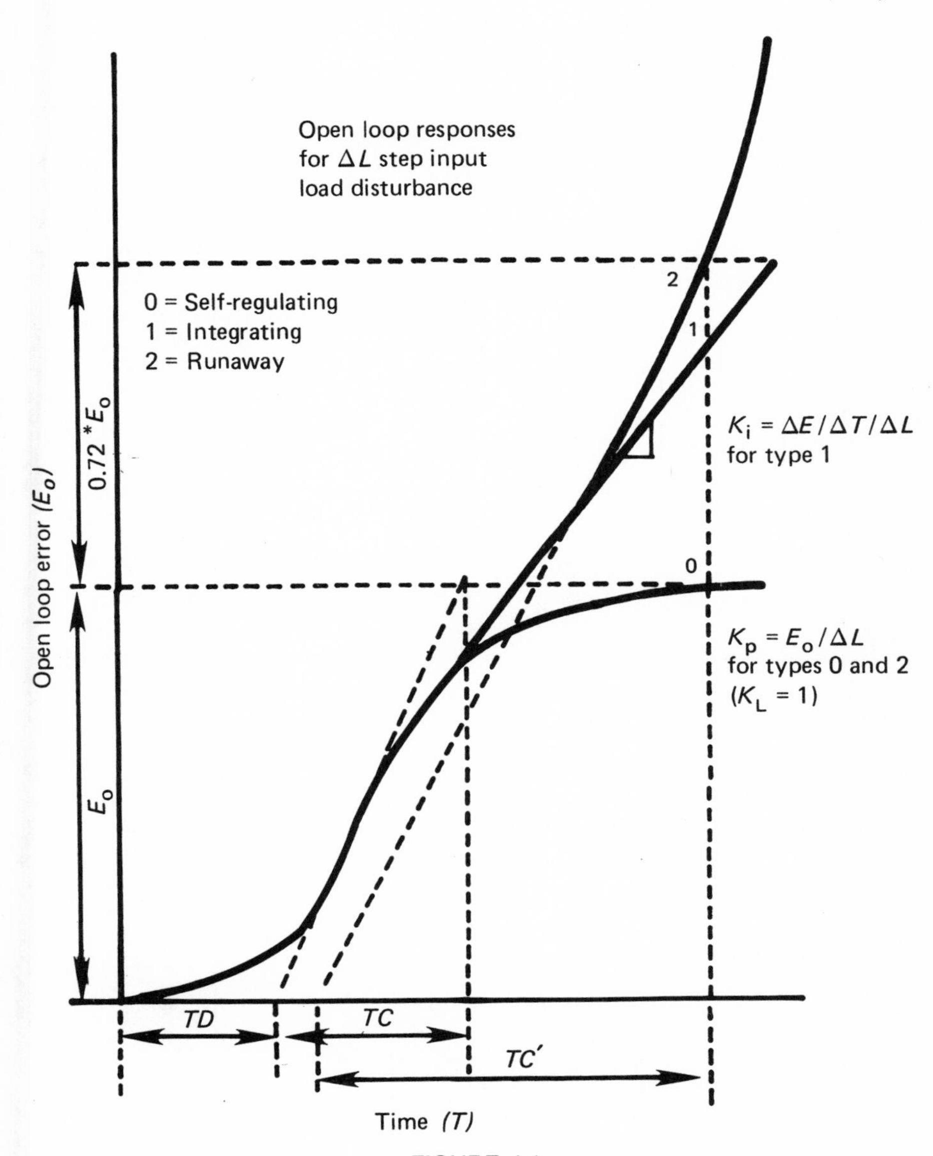


FIGURE 4.1
Graphical method to estimate dead time and time constant.

TABLE 4.1
Process Dead Times and Time Constants

| *Equipment Type* | *Controlled Variable* | *Manipulated Variable* | *Equations for Dead Times and Time Constants* | *Reference* |
|---|---|---|---|---|
| Agitated vessel | Concentration | Reagent | $TD = \frac{V}{F + F_a}$ <br> $TC = \frac{V}{F} - TD$ | Shinskey, 1979 |
| Static mixer | Concentration | Reagent | $TD = 0.75 * \frac{V}{F}$ <br> $TC = 0.25 * \frac{V}{F}$ | Bor, 1971 |
| Pipeline | Liquid flow | Pump speed | $TC = \frac{L_p * A_p * C_p^2}{F}$ | Shinskey, 1979 |
| Pipeline | Liquid flow | Valve | $TC = \frac{D_p}{4 * C_p * V_f * (1 + R)}$ | Harriot, 1964 |
| Heat exchanger | Tube temperature | Shell temperature | $TC_t = \frac{M_t * C_t / N}{H_t * A_t / N + W_t * C_t}$ <br> $TC_w = \frac{M_w * C_w}{H_t * A_t + H_s * A_s}$ <br> $TC_s = \frac{M_s * C_s / N}{H_s * A_s / N + W_s * C_s}$ | Roffel, 1982 |
| Exothermic reactor | Reactants temperature | Coolant temperature | $TC' = \frac{M_r * C_r}{UA + W_r * C_r - \Delta Q / \Delta T}$ <br> $TC = \frac{M_c * C_c}{UA}$ <br> $TD = \frac{M_c}{W_c}$ | Harriot, 1964 |

| | | | | |
|---|---|---|---|---|
| Process heater | Tube temperature | Radiation heat | $TC = \frac{M_t{*}C_t{*}M_w{*}C_w}{M_t{*}C_t + M_w{*}C_w} * \frac{1}{A_t{*}H_t}$ | Roffel, 1982 |
| Furnace | Gas pressure | Gas flow | $TC = \frac{V}{P_a * \Sigma 1/R_j}$ <br> $R_j = \frac{2{*}\Delta P_j}{F_j}$ (turbulent flow) <br> $R_j = \frac{P_j}{F_j}$ (sonic flow) | Harriot, 1964 |
| Biological reactor | Cell concentration | Nutrient dilution | $TC' = 1/U$ <br> $U = U_x * \frac{C_i}{K_i + C_i} - K_e$ | Bailey, 1977 |

Nomenclature (in order of appearance):

$TC$ = time constant (negative feedback)
$TD$ = dead time
$TC'$ = time constant (positive feedback)
$V$ = vessel, static mixer, or furance volume
$F$ = volumetric throughput flow
$F_a$ = agitator pumping rate (volumetric flow)
$L_p$ = length of pipeline
$A_p$ = cross-sectional area of pipeline
$C_p$ = flow coefficient or friction factor for pipeline
$D_p$ = diameter of pipeline
$V_f$ = fluid velocity in pipeline
$R$ = ratio of control valve to pipeline pressure drop
$M_t$ = mass of fluid in heat exchanger tubes
$C_t$ = specific heat of fluid in heat exchanger tubes
$H_t$ = heat transfer coefficient of heat exchanger tubes
$A_t$ = heat transfer area of heat exchanger tubes
$W_t$ = mass flow of fluid in heat exchanger tubes
$N$ = number of segments in heat exchanger
$M_w$ = mass of tube walls in heat exchanger
$C_w$ = specific heat of tube walls in heat exchanger
$M_s$ = mass of fluid in heat exchanger shell
$C_s$ = specific heat of fluid in heat exchanger shell
$H_s$ = heat transfer coefficient of heat exchanger shell
$A_s$ = heat transfer area of heat exchanger shell
$W_s$ = mass flow of fluid in heat exchanger shell
$M_r$ = mass of reactants
$C_r$ = specific heat of reactants
$W_r$ = mass flow of reactants
$M_c$ = mass of coolant
$C_c$ = specific heat of coolants
$W_c$ = mass flow of coolant
$\Delta Q$ = change in heat of reaction
$\Delta T$ = change in reactant temperature
$P_a$ = atmospheric pressure
$R_j$ = resistance of connection $j$ to furnace
$\Delta P_j$ = pressure drop across resistance $j$
$F_j$ = volumetric flow through resistance $j$
$P_j$ = inlet pressure at resistance $j$
$U$ = generation rate of cells
$U_x$ = maximum generation rate of cells
$C_i$ = concentration of nutrient
$K_i$ = concentration of nutrient when $U = U_x/2$
$K_e$ = substrate consumption rate for cell survival

summation of each small time constant multiplied by its respective $Y$ factor. Equation 4.1 was documented by Ziegler and Nichols in a reply to a question on their paper (Ziegler & Nichols, 1943).

$$Y_i = 1 + \frac{1}{X_i} - \frac{1}{X_i^Z} - \frac{\ln(X_i)}{1 - X} \qquad (4.1)$$

$$TD_i = \Sigma Y_i{}^*TC1_i \qquad (4.2)$$

where

$Y_i$ = effective dead time factor

$X_i$ = ratio of small to large time constant $\left[X_i = \frac{TC1_i}{TC2}\right]$

$\ln(X_i)$ = natural logarithm of $X_i$

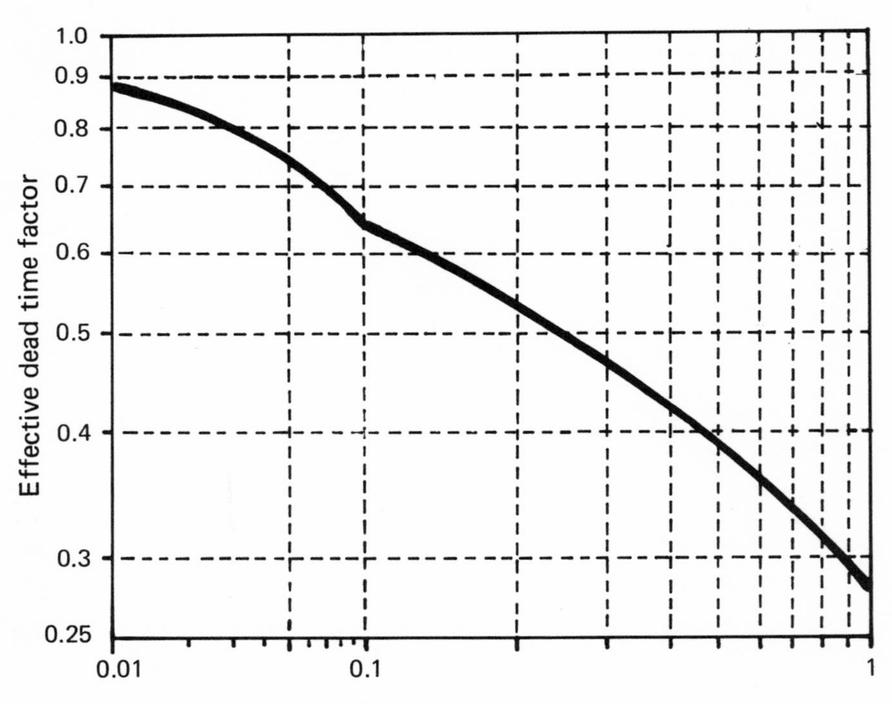


FIGURE 4.2
Effective dead time factor for small and large time constants.

$$Z = \text{exponent}\left[Z = \frac{1}{1 - X_i}\right]$$
$TC1_i$ = small time constant
$TC2$ = large time constant
$TD$ = effective dead time (time delay)

If the process consists of several equal large time constants in addition to the smaller ones, the summation of the large time constants should be used as *TC*2 in the computation of ratio *X* for Equation 4.1 and Figure 4.2. The effective time constant associated with the effective dead time is *TC*2 minus the effective dead time divided by a correction factor. The correction factor is one plus the *Y* factor from Figure 4.3. The FORTRAN

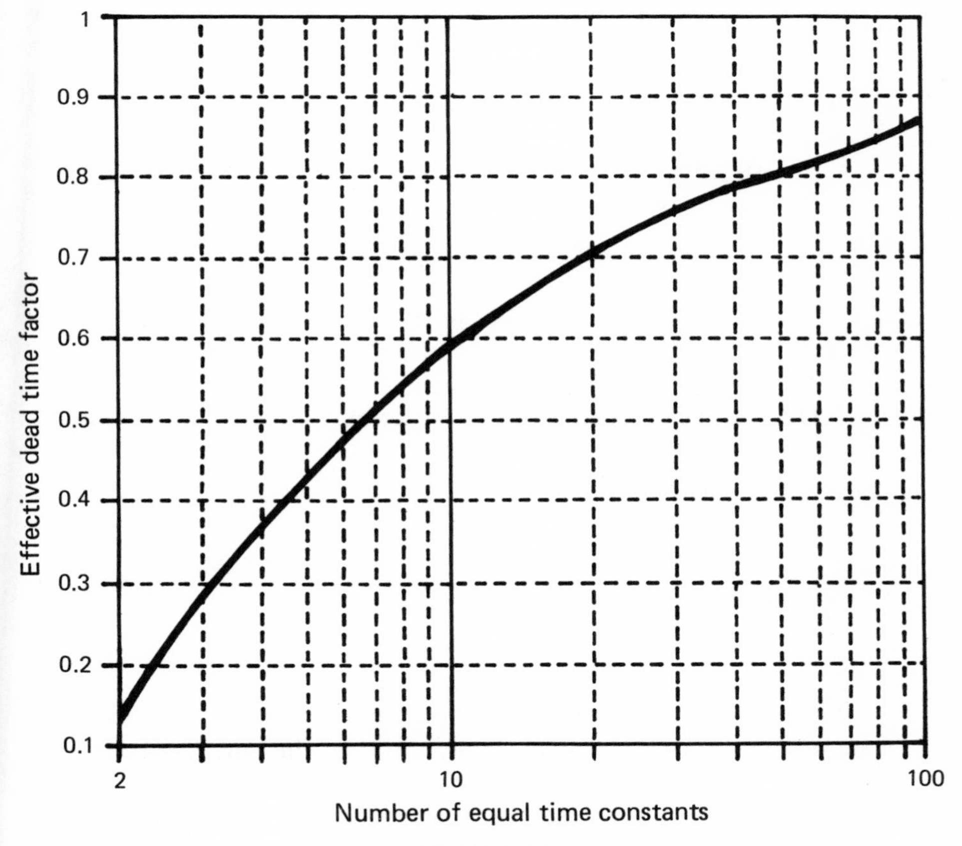


FIGURE 4.3
Effective dead time factor for several large equal time constants.

subroutine in Appendix J can be used to compute the effective dead time and the effective time constant. Previous applications of the Ziegler-Nichols tuning rules to loops with equal time constants yielded excessively oscillatory closed loop responses because the proportional band was too low due to the fact that the uncorrected effective time constant was too large. The effective dead time is the summation of each small time constant multiplied by its respective $Y$ factor from Equation 4.1 or Figure 4.2 plus the summation of the large equal time constants multiplied by their $Y$ factor from Equation 4.3 or Figure 4.3 (see Example 4.4.1 for details on application of this technique). Equation 4.3 is the result of linear regression to fit a polynomial with transforms to the plot documented by Ziegler and Nichols (Ziegler & Nichols, 1943).

$$Y = A_0 + A_1*[\ln(N)]^1 + A_2*[\ln(N)]^2 + A_3*[\ln(N)]^3 + A_4*[\ln(N)]^4 \quad (4.3)$$

$$TD = Y*N*TC \quad (4.4)$$

where

$Y$ = effective dead-time factor
$N$ = number of equal large time constants in series
$TC$ = each large time constant
$TD$ = effective dead time (time delay)
$\ln(N)$ = natural logarithm of $N$
$A_0 = -0.06695235$
$A_1 = 0.2595017$
$A_2 = 0.08436275$
$A_3 = -0.04161446$
$A_4 = 0.004497025$

If the process consists of multiple pieces of equipment each with a time constant and each affected by the operating conditions of adjacent pieces of equipment, the time constants are called interactive. The equivalent noninteractive time constants must be calculated via the equations in Appendix F before

Figure 4.2 or 4.3 can be used. The equivalent noninteractive time constants will show a larger spread in values so that the resulting dead time will be smaller and the resulting time constant will be larger. The proportional band and the integral time will be smaller and consequently the accumulated and peak errors will be smaller. For example, two 1.0-minute interactive time constants in series would have equivalent noninteractive time constants of 2.0 minutes and 0.66 minute. A gas pressure control system that has a series of tanks or sections of pipeline has interacting time constants. If the number of tanks or sections of pipelines is very large, the system approaches what is called a ''distributed'' system. The response of a distributed system at its midpoint can be approximated as a dead time equal to one-tenth the sum of the interactive time constants and a single time constant equal to one-half the sum of the interactive time constants. Dead time can also originate from transportation delay in process equipment and analyzer sample lines or from time sampling devices in instruments such as chromatographs or digital controllers. For unagitated equipment and sample lines, the effective dead time is nearly equal to the volume divided by the throughput. For chromatographs where the calculation is done at the end of the sample period, the dead time is equal to 1.5 times the sample period. For digital controllers where the calculation is done at the beginning of the sample period, the dead time is equal to 0.5 times the sample period (see Sections 5.4 and 6.2 for more details).

Equations to estimate the time constants of various types of processes and references for more information are summarized in Table 4.1.

## 4.2 *Accuracy of Dead Time and Time Constant Approximation*

S. G. Lloyd and G. D. Anderson showed with Bode plots that a dead time plus time constant approximation predicted with negligible error the same natural frequency of oscillation and associated dynamic gain, and thus controller mode settings, as four equal time constants in series (Lloyd & Anderson, 1971). E. H. Bristol showed that a two time constant approximation to an eight time constant process that had less than a 3 percent error in matching the time response caused instability when used to tune the controller (the predicted controller gain was more than three times too large) because dead time was omitted from the approximation (Bristol, 1970). The Pade approximation of dead time as a lead–lag can result in inverse response. Such multiple time constant approximations of process dynamics require significantly more effort and produce erroneous results for small errors. The dead time plus time constant approximation is simple and safe for self-regulating processes (the inaccuracy is small and in the direction of predicting a slightly lower controller gain and therefore larger accumulated and peak error). The dead time plus time constant approximation will be used in this section to describe a self-regulating process. For non-self-regulating processes, an integrator or positive feedback time constant will be added to the approximation.

## *4.3*
## *Self-Regulating Processes*

The open loop response of a self-regulating process reaches a final steady-state value for a step disturbance in load or controller output. Flow loops, level loops that do not have discharge pumps (liquid head sets discharge flow), pressure loops that have a discharge opening whose ratio of volume to throughput is small, temperature loops that do not control exothermic reactions, oxidation reduction potential (ORP) loops, and pH loops are self-regulating. Continuous concentration control loops are self-regulating whereas batch concentration control loops are integrating. Batch ORP or pH loops are integrating but appear to be self-regulating because of the dramatic decrease in the slope of the ORP or pH versus concentration curve at the extremes of the measurement range. The self-regulating process is adequately represented by a steady-state gain $K_p$, which, multiplied by the disturbance size, gives the change in steady-state value of the process (the open loop error—see Equation 3.3) and by a dead time plus time constant approximation that gives the time response of the process (see Figure 4.1). Simulation results and Bode plot analysis show that the ultimate period of oscillation can be approximated by Equation 4.5 for self-regulating processes. (Equation 4.5 will predict ultimate periods approximately 10 percent larger than an exact trial-and-error solution for dead time to time constant ratios larger than 5.0.)

$$T_u = 2*\left[1 + \left[\frac{TC}{(TC + TD)}\right]^{0.65}\right]*TD \tag{4.5}$$

where

$T_u$ = ultimate period of oscillation
$TC$ = time constant
$TD$ = dead time

Equation 4.5 shows that the ultimate period approaches twice the dead time if the dead time is much larger than the time constant. This occurs rarely in chemical loops except when a chromatograph or laboratory analysis is used for the measurement or when the change in gain with error is extremely large (see Example 9.4). Equation 4.5 also shows that the ultimate period approaches four times the dead time if the dead time is much smaller than the time constant, which is usually the case for chemical loops.

The ultimate period, $T_u$, is the period of the natural frequency of the loop. If the loop controller's reset and rate action are turned off and the proportional band is decreased (gain is increased) until the loop oscillates with constant amplitude, the period of these oscillations is the ultimate period, $T_u$. Ziegler and Nichols (Ziegler & Nichols, 1942) showed that the controller mode settings could be estimated from the proportional band, $PB_u$, for which sustained constant amplitude oscillations occurred and from the ultimate period, $T_u$.

For a proportional plus reset (PI) controller:

$$PB = 2.22*PB_u \tag{4.6}$$

$$T_i = 0.83*T_u \tag{4.7}$$

For a proportional plus reset plus rate (PID) controller:

$$PB = 1.67*PB_u \tag{4.8}$$

$$T_i = 0.50*T_u \tag{4.9}$$

$$T_d = 0.125*T_u \tag{4.10}$$

where

$PB$ = proportional band of the controller (100 percent/gain)
$PB_u$ = ultimate proportional band to generate oscillations
$T_u$ = ultimate period of the oscillations (minutes)
$T_i$ = reset (integral) time of the controller (minutes/repeat)
$T_d$ = rate (derivative) time of the controller (minutes)

The foregoing equations show that the addition of rate (derivative) action to a controller allows a smaller proportional band and smaller reset (integral) time, which means a smaller accumulated and peak error. The amount of rate allowable for a given dead time decreases as the time constant decreases since the ultimate period decreases from an upper limit of 4* $TD$ to a lower limit of 2*$TD$. The ratio of integral time to rate time should be equal to or less than 4:1 for the "ideal" controller (see Chapter 5). The actual permissible amount of derivative depends on the amount of noise. Loops with small time constants tend to be noisier since the attenuation of the noise by filtering is inversely proportional to the time constant (attenuation $= \dfrac{T_u}{2\pi*TC} < 1$). The proportional band can be estimated from the ultimate period, time constant, and steady-state gains of the loop by use of Equation 4.11.

For $\frac{T_u}{TC} < 4$:

$$PB = \frac{K_g*100*T_u*K_v*K_p*K_m}{2\pi*TC} \quad (4.11)$$

For $\frac{T_u}{TC} > 4$:

$$PB = K_g*100*K_v*K_p*K_m \quad (4.12)$$

where

$PB$ = proportional band of the controller (100 percent/gain)
$K_g$ = gain factor for the controller (1.5 typically)
$T_u$ = ultimate period of oscillation (minutes)
$TC$ = time constant (minutes)
$K_v$ = steady-state valve gain
$K_p$ = steady-state process gain
$K_m$ = steady-state measurement gain

The $K_g$ factor used in the proportional band and error calculations depends on the manufacturer and type of controller, for example:

$K_g = 1*K_u$ for analog controllers
$K_g = 1.25*K_u$ for digital controllers
$K_u = 2.0$ (PI), $K_u = 1.5$ (PID)

Since the ultimate period, $T_u$, is equal to approximately four times the dead time for most self-regulating chemical processes, the proportional band is proportional to the ratio of dead time to time constant ($TD/TC$), which results in the equation used by Shinskey (Shinskey, 1979). The peak error is then also proportional to this ratio and the accumulated error is proportional to the dead time squared divided by the time constant.

Substitute Equations 4.9 and 4.11 into 3.4 and 3.10.

For $\frac{T_u}{TC} < 4$:

$$E_i = \left[\frac{K_g*T_u}{2\pi*TC}\right]*0.5*T_u*E_o \tag{4.13}$$

$$E_x = \left[\frac{1.1*K_g*T_u}{2\pi*TC}\right]*E_o \tag{4.14}$$

where

$E_i$ = accumulated (integrated) error
$E_x$ = peak (maximum) error
$E_o$ = open-loop steady-state error

For small $TD/TC$ ratios, the peak error is approximately equal to $(TD/TC)*E_o$ and the accumulated error is approximately equal to $(TD/TC)*2*TD*E_o$ since $T_u = 4*TD$. Thus, if the dead time increases by a factor of 10, the peak error increases by a factor of 10 and the accumulated error increases by a factor of 100. Further study of the effect of dead time leads to some additional conclusions. First, if the dead time is zero, control is perfect (the accumulated and peak errors are zero). Thus control is perfect for a loop with a single time constant. However such a loop does not exist because even if the process has only one time constant, the loop valve and transmitter add time constants and dead times (for this case where the valve and transmitter dynamics limit the performance, it is important to make them as fast as possible). Second, the largest time constant should be made as large as possible and the other time constants as small as possible, especially the larger of the remaining ones (see Figure 4.1). Third, transportation delays and sampling periods should be made appreciably smaller than the next largest time constant to ensure that these delays and periods do not limit the loop performance.

Figures 4.4 and 4.5 are three-dimensional plots of Equations 4.13 and 4.14 for the accumulated and peak errors ratioed to the open loop error. Note that as the loop dead time approaches zero, the errors approach zero.

## 4.4 *Integrating Processes*

The open loop response of an integrating process does not reach a final steady-state value after a step disturbance in load or controller output but continues to ramp off until a physical limit is reached, such as tank overflow for a level loop and equipment rupture for a pressure loop. Level loops that have discharge pumps or valve on flow in, pressure loops that have

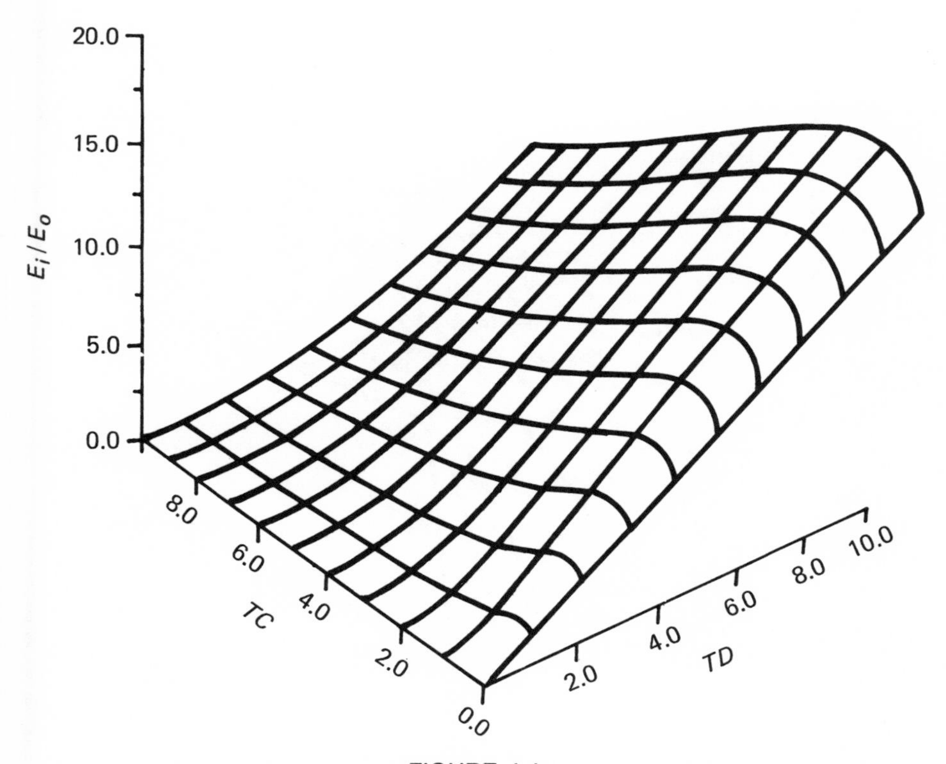


FIGURE 4.4
Three-dimensional accumulated error for a self-regulating process.

no discharge opening or whose ratio of volume to throughput is large, and conductivity loops that control total dissolved solids are integrating. This type of process is adequately represented by a steady gain, $K_p$, which, when multiplied by the disturbance size, gives the integrator input; by a dead time plus time constant approximation for the self-regulating part of the process, and by an integrator gain that is the slope of the ramp of the process for a step disturbance. Simulation results and Bode plot analysis show that the ultimate period of oscillation can be approximated by Equation 4.15 for integrating processes:

$$T_u = 4*\left[1 + \left[\frac{TC}{TD}\right]^{0.65}\right]*TD \qquad (4.15)$$

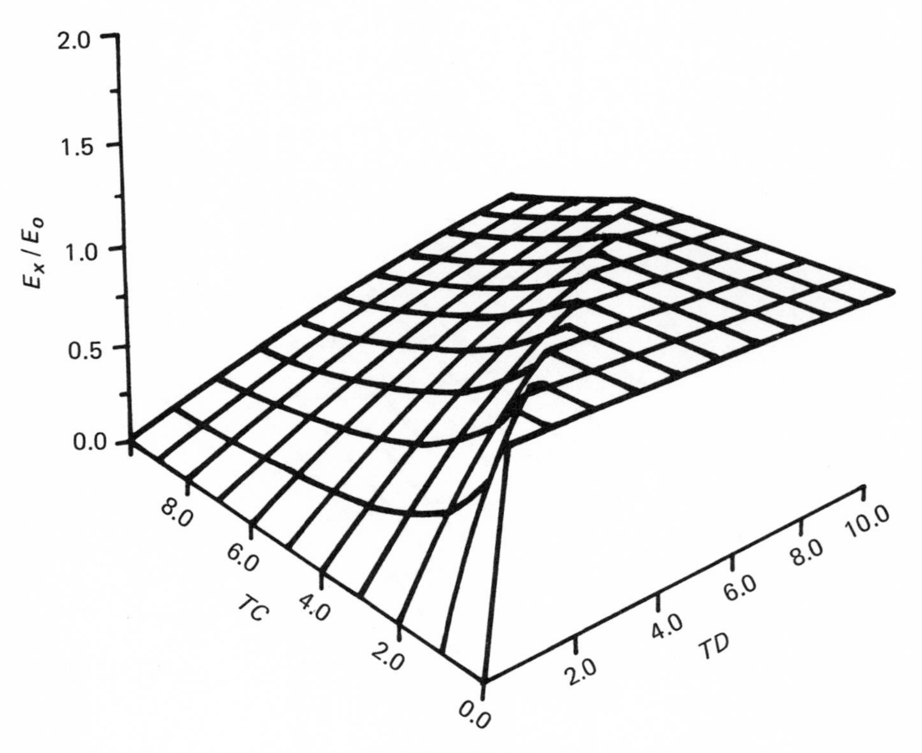


FIGURE 4.5
Three-dimensional peak error for a self-regulating process.

where

$T_u$ = ultimate period of oscillation
$TC$ = time constant of the self-regulating part of process
$TD$ = dead time of the self-regulating part of process

Equation 4.15 shows that the ultimate period is four times the dead time, if the time constant is extremely small compared with the dead time. This is the relationship that Shinskey uses for integrating processes. However the time constant is normally large compared with the dead time so that the ultimate period is much larger for an integrating process than for a self-regulating process with the same dead time and time constant. For example, if the dead time is 0.1 minute and the time constant is 1 minute, the ultimate period for a self-regulating process is 3.88 times the dead time or 0.388 minute whereas the ultimate period for an integrating process is 21.86 times the dead time or 2.186 minutes (56 times larger). Consequently the reset time (minutes per repeat) and rate time settings are also 56 times larger. These conclusions agree with the general observation that integrating processes require much less reset action (much more reset time) and much more rate action than self-regulating processes.

The proportional band can be estimated from the ultimate period, time constant, steady-state gains, and integrator gain of the loop.

For $\frac{T_u}{TC} < 4$:

$$PB = \frac{K_g*100*T_u^2*K_v*K_p*K_m}{(2\pi)^2*TC} \tag{4.16}$$

For $\frac{T_u}{TC} > 4$:

$$PB = \frac{K_g*100*T_u*K_v*K_p*K_m}{2\pi} \tag{4.17}$$

(The process gain, $K_p$, must include the integrator gain, $K_i$, so that the inverse time units of $K_i$ cancel the time units of the extra $T_u$ term.)

Except for an extremely small time constant as compared with the dead time (Shinskey's case), the proportional band is generally smaller for an integrating process than for a self-regulating process with the same time constants and dead time based on Equation 4.16 since the integrator gain $K_i$ is small ($K_i << 1.0$ so that $T_u{}^*K_i << 2\pi$). The small proportional band results in small offsets, which is one reason why reset action is not used in level control loops. If level measurement noise due to turbulence or bubblers necessitates the widening of the proportional band, reset action may be added to eliminate the resulting offset. Simulation program results show that if the proportional band is 100 times larger than optimum, the oscillation goes from quarter-amplitude damping to nearly undamped (sustained oscillations), with a period 25 times longer than the ultimate period $T_u$.

Even though this case seems extreme, the proportional bands required of some integrating processes are so small that they are below the lower limit of the controller. This situation may result in the use of proportional bands that approach the upper stability limit. If the ultimate oscillation method is used, the nearly undamped oscillations may be confused with the oscillations associated with the ultimate proportional band and attempts made to increase the proportional band further will only aggravate the problem. This window of allowable proportional bands has not been discussed in the literature to date, although such a window has been documented for runaway processes such as exothermic reactions (see Chapter 4).

The ramping time response of an integrating process to a step disturbance can be viewed as the time response for an infinite time constant. In fact, a process with an extremely large time constant compared with dead time emulates an integrating process. The peak error is so small since the ratio $TD/TC$ is so small that the time response looks like a ramp in the control region. The integrator gain, $K_i$, is equal to the slope of the tangent drawn at the beginning of the exponential response curve. Since the control region is near the start of the exponential response, the tangent represents the ramp response of the pseudointegrator. This pseudointegrator also has a window of allowable proportional bands just as the real integrator does. Also Equations 4.15 through 4.17 can be used if the next largest time constant is used for $TC$. Many furnace pressure control loops have pseudointegrators since the volume to throughput ratio is large so that the $TC$ is large and valve and measurement dynamics determine the dead time so that the $TD$ is small. Control ranges of fractional inches of water are achievable only because the $TD/TC$ ratio is extremely small for these furnace pressure control loops.

$$K_i = \frac{K_p}{TC} \tag{4.18}$$

where

$K_i$ = gain of the pseudointegrator (slope of ramp response)
$K_p$ = steady-state process gain associated with $TC$
$TC$ = extremely large time constant compared with the dead time

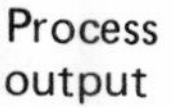

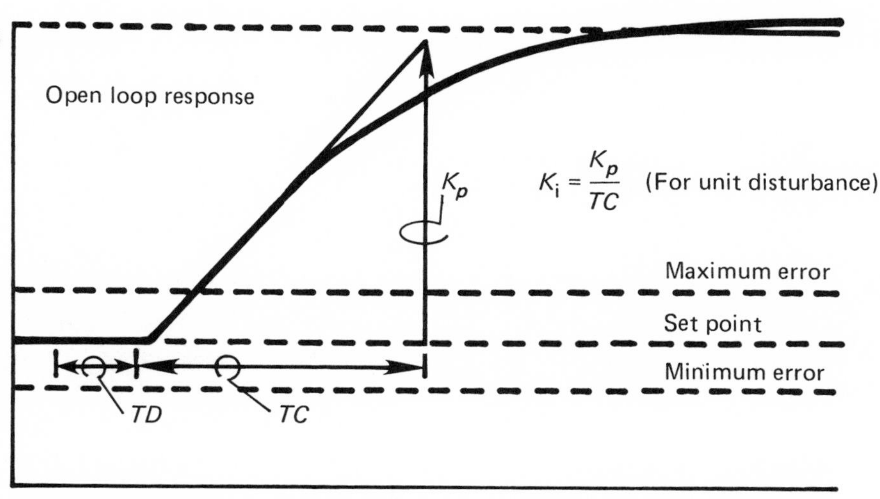


FIGURE 4.6
Pseudointegrator gain approximation.

For an integrating process with $T_u/TC < 4$, the peak error is proportional to the ultimate period squared divided by the time constant and the accumulated error is proportional to the ultimate period cubed divided by the time constant.

Substitute Equations 4.9 and 4.16 into 3.4 and 3.10:

For $\frac{T_u}{TC} < 4$:

$$E_i = \left[\frac{K_g{}^*T_u^2}{(2\pi)^{2*}TC}\right]{}^*0.5{}^*T_u{}^*E_o \tag{4.19}$$

$$E_x = \left[\frac{1.1{}^*K_g{}^*T_u^2}{(2\pi)^{2*}TC}\right]{}^*E_o \tag{4.20}$$

where

$E_i$ = accumulated (integrated) error
$E_x$ = peak (maximum) error
$E_o$ = open-loop steady-state error (see Equation 3.3)

(The process gain, $K_p$, must include the integrator gain, $K_i$, in the calculation of the open loop error, $E_o$, so that the inverse time units of $K_i$ cancel the time units of the extra $T_u$ term in Equations 4.19 and 4.20.)

Batch processes (integrating) will have approximately the same accumulated and peak errors as continuous processes (self-regulating) if the tanks both have large volumes, low feed flows, and high agitation and the transmitter and valve dynamics are fast so that both types of loops have an ultimate period equal to four times the dead time. If any of the aforementioned conditions do not exist, the accumulated and peak errors of the batch process will be larger. However, if there is sufficient time and working volume, the batch processing may continue until the error is reduced to within specifications. For example, doses of acidic and basic reagents can be added in smaller and smaller increments to a large holding tank for pH adjustment before transfer to downstream equipment.

Figures 4.7 and 4.8 are three-dimensional plots of Equations 4.19 and 4.20 for the accumulated and peak errors ratioed to the open loop error. Note that as the loop dead time approaches zero, the errors approach zero.

## *4.5 Runaway Processes*

The open loop response of a runaway (positive feedback) process does not reach a steady-state value after a step disturbance in load or controller output but increases at a rate that

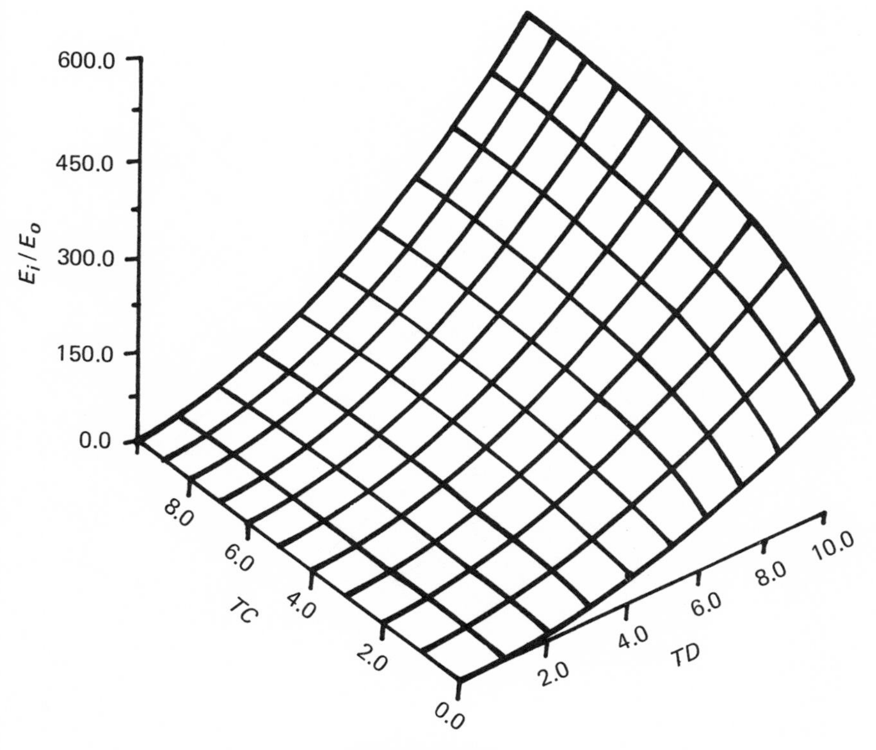


FIGURE 4.7
Three-dimensional accumulated error for an integrating process.

increases with time (slope of response curve increases with time) until a physical limit is reached such as vessel rupture or reactant depletion for exothermic reactions and overspeed damage for dynamic compressors. Temperature loops that control exothermic reactions where the slope of the heat removal curve is less than the slope of the heat generation curve, concentration loops that control biologic reactions where the slope of the dilution curve is less than the slope of the cell generation curve, and some speed loops that control dynamic compressors during surge where the rotor inertia is low have runaway processes. The positive feedback is in the process itself. The cause-and-effect relationship is illustrated as follows for three types of processes.

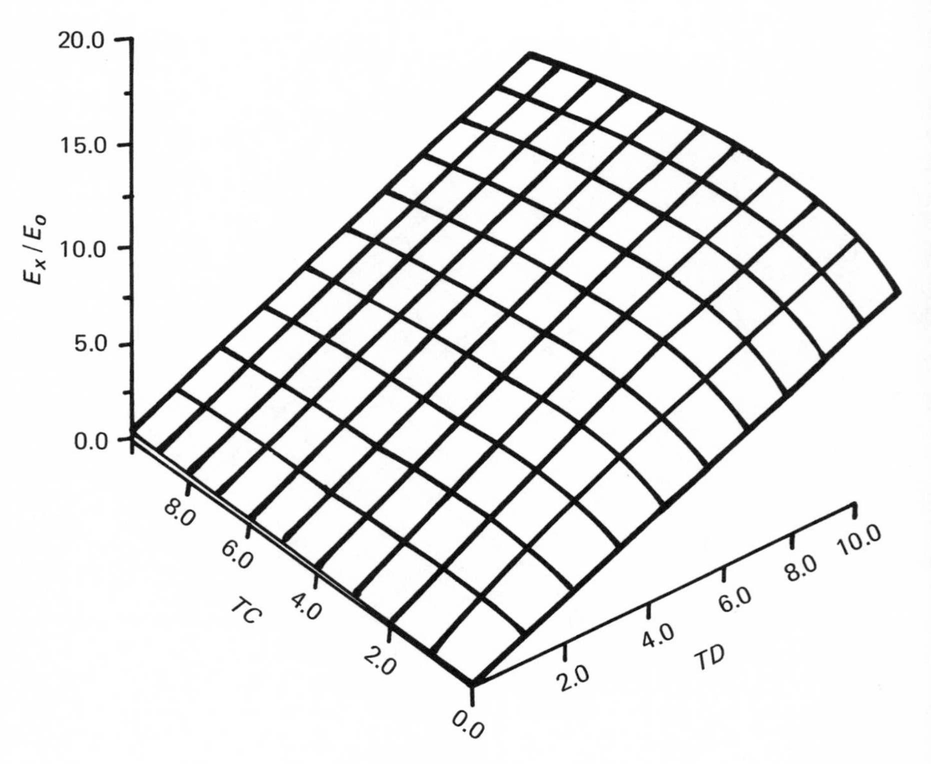


FIGURE 4.8
Three-dimensional peak error for an integrating process.

*Exothermic reactor:*

1. Increase in reaction temperature
2. Increase in reaction rate
3. Increase in heat generation
4. Increase in reaction temperature

*Biological reactor:*

1. Increase in cell concentration
2. Increase in cell generation rate
3. Increase in cell concentration

*Dynamic compressor in surge:*

1. Increase in speed of impeller
2. Increase in unloading of impeller
3. Increase in speed of impeller

This type of process is adequately represented by a steady-state gain, $K_p$, which, when multiplied by the disturbance size, gives the positive feedback input; by a dead time plus time constant approximation for the self-regulating parts of the process; and by a positive feedback time constant, $TC'$ ($TC'$ is the time required for the positive feedback part of the process to reach 1.72818 times its input). Simulation results and Nyquist plot analysis show that the ultimate period of oscillation can be approximated by Equation 4.21 for runaway (positive feedback) processes.

$$T_u = 4*\left[1 + \left[\frac{N}{D}\right]^{0.65}\right]*TD \tag{4.21}$$

where

$N = (TC' + TC)*TC'*TC$
$D = (TC' - TC)*(TC' - TD)*TD$
$T_u$ = ultimate period of oscillation
$TC$ = time constant of the self-regulating part of the process
$TD$ = dead time of the self-regulating part of the process
$TC'$ = time constant of the positive feedback part of the process

Equation 4.21 reduces to Equation 4.15 for the integrating process if *TC′* is much larger than *TC* and *TD*; this is as expected since for very large values of *TC′* the time response approaches a ramp in the control region. Equation 4.21 shows that if *TC* or *TD* equals *TC′*, the period becomes infinitely large. Nyquist plot analysis confirms that if *TC* or *TD* is equal to or greater than *TC′*, the loop is unstable regardless of controller mode settings. The minimum and maximum allowable proportional bands approach each other until they overlap as *TC* or *TD* approaches *TC′* in size. The window of allowable proportional is closed. Figures 4.9 and 4.10 show the effect of various values of dead time, *TD*, and self-regulating time constant, *TC*, on the size of the window. The concept of too small a proportional

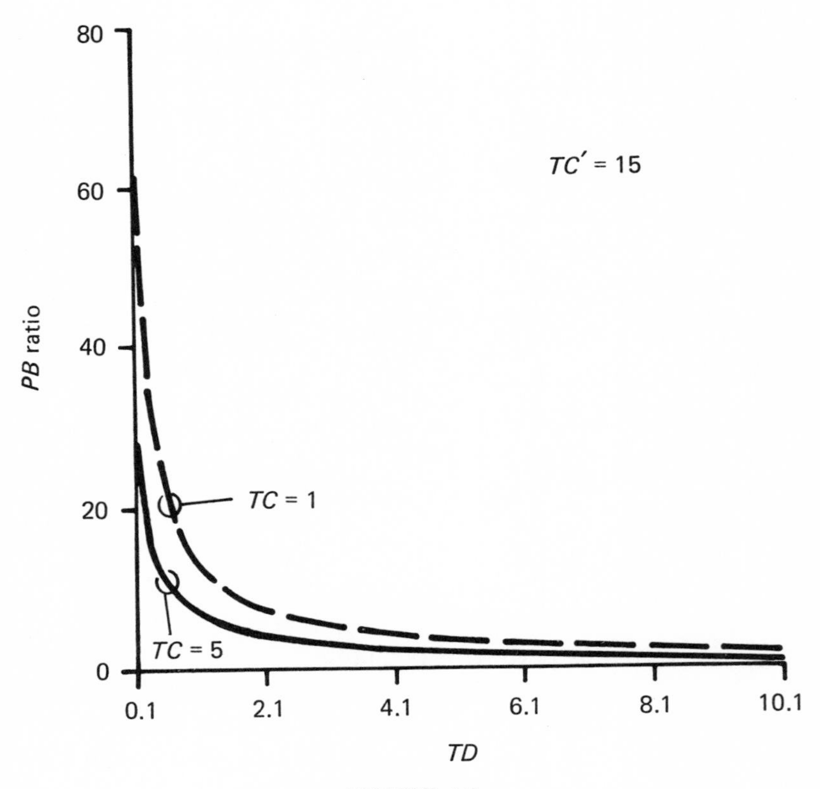


FIGURE 4.9
Effect of dead time on proportional band window.

band (too large a gain) is familiar but the concept of too large a proportional band (too small a gain) causing instability merits some discussion. Too large a proportional band in the temperature controller of an exothermic reaction will result in a small temperature disturbance growing uncorrected and running away until a physical limit is reached or the other modes (reset or rate) overcompensate and start unstable oscillations. W. L. Luyben developed graphs from Nyquist plots (for $TD = 0$) which show the ratio of maximum to minimum allowable controller gain as a function of vessel size, heat transfer coefficient times heat transfer area $U^*A$, and temperature measurement time constant $TC_m$ for a proportional-only controller (Luyben & Cheung, 1978).

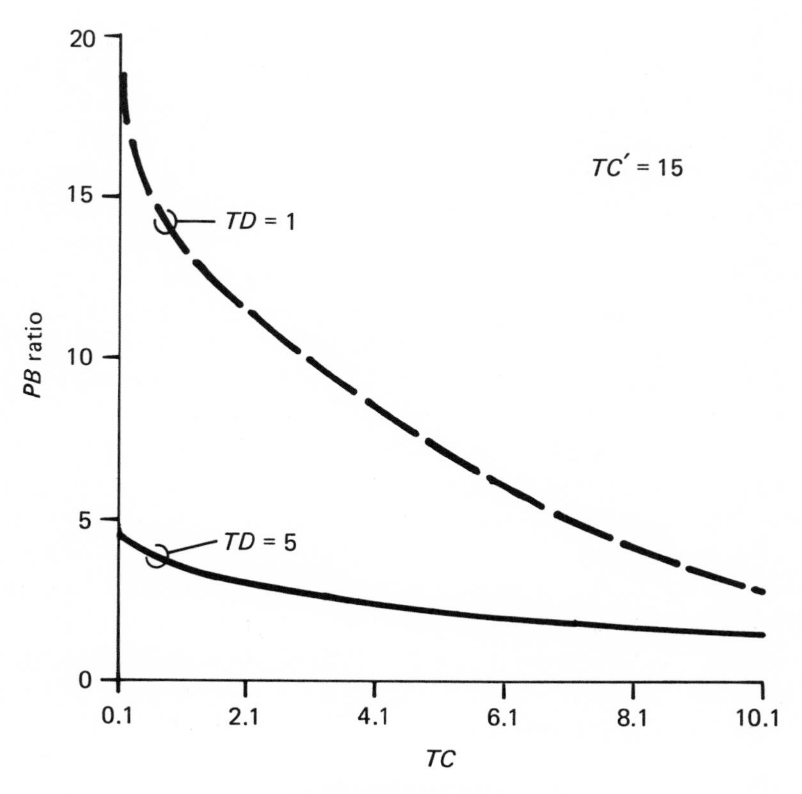


FIGURE 4.10
Effect of time constant on proportional band window.

The proportional band can be estimated from the ultimate period, time constant $TC$, steady-state gains, and positive feedback time constant $TC'$ of the loop.

For $\frac{T_u}{TC} < 4$:

$$PB = \frac{K_g*100*T_u^2*K_v*K_p*K_m}{(2\pi)*TC*TC'} \tag{4.22}$$

For $\frac{T_u}{TC} > 4$:

$$PB = \frac{K_g*100*T_u*K_v*K_p*K_m}{2\pi*TC'} \tag{4.23}$$

Equations 4.22 and 4.23 are equivalent to 4.16 and 4.17 if $K_i = 1/TC'$ and $TC'$ is much larger than $TC$ or $TD$ so that the $T_u$'s are equal. The runaway (positive feedback) process has a much larger period and smaller proportional band than the self-regulating process with the same $TC$ and $TD$. The runaway process has a slightly larger period and smaller proportional band than the integrating process with the same $TC$ and $TD$. The optimum proportional band may be below the lower limit and the optimum reset and rate times above the upper limit of conventional analog controllers. Many runaway processes operate best with proportional plus derivative controllers because the reset action required is so small because of the large ultimate period that the reset can be turned off (it is better to have too little reset action than too much).

For a runaway process with $T_u/TC < 4$, the peak error is proportional to the ultimate period squared divided by both the time constants ($TC$ and $TC'$) and the accumulated error is proportional to the ultimate period cubed divided by the time constants.

Substitute Equations 4.9 and 4.22 into 3.4 and 3.10.

For $\frac{T_u}{TC} < 4$:

$$E_i = \left[\frac{K_g{*}T_u^2}{(2\pi)^2{*}TC{*}TC'}\right]{*}0.5{*}T_u{*}E_o \tag{4.24}$$

$$E_x = \left[\frac{1.1{*}K_g{*}T_u^2}{(2\pi)^2{*}TC{*}TC'}\right]{*}\mathrm{E}_o \tag{4.25}$$

where

$E_i$ = accumulated (integrated) error
$E_x$ = peak (maximum) error
$E_o$ = opem-loop steady-state error (see Equation 3.3)

Figures 4.11 and 4.12 are three-dimensional plots of Equations 4.24 and 4.25 for the accumulated and peak errors ratioed

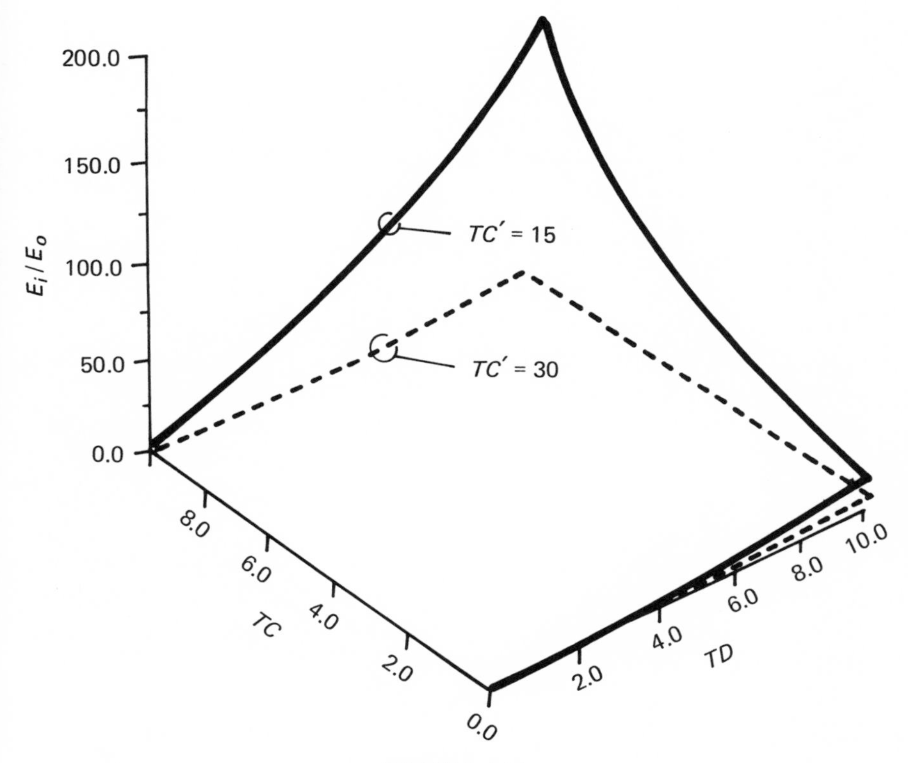


FIGURE 4.11
Three-dimensional accumulated error for a runaway process.

to the open loop error. Note that as the loop dead time approaches zero, the errors approach zero.

The minimum and maximum proportional bands for a runaway process can be estimated as follows:

$$PB_{\max} = 80 * K_v * K_p * K_m \tag{4.26}$$

$$PB_{\min} = 0.5 * PB \tag{4.27}$$

where

$PB_{\max}$ = maximum proportional band for stability
$PB_{\min}$ = minimum proportional band for stability
$PB$ = proportional band from Equation 4.22 or 4.23

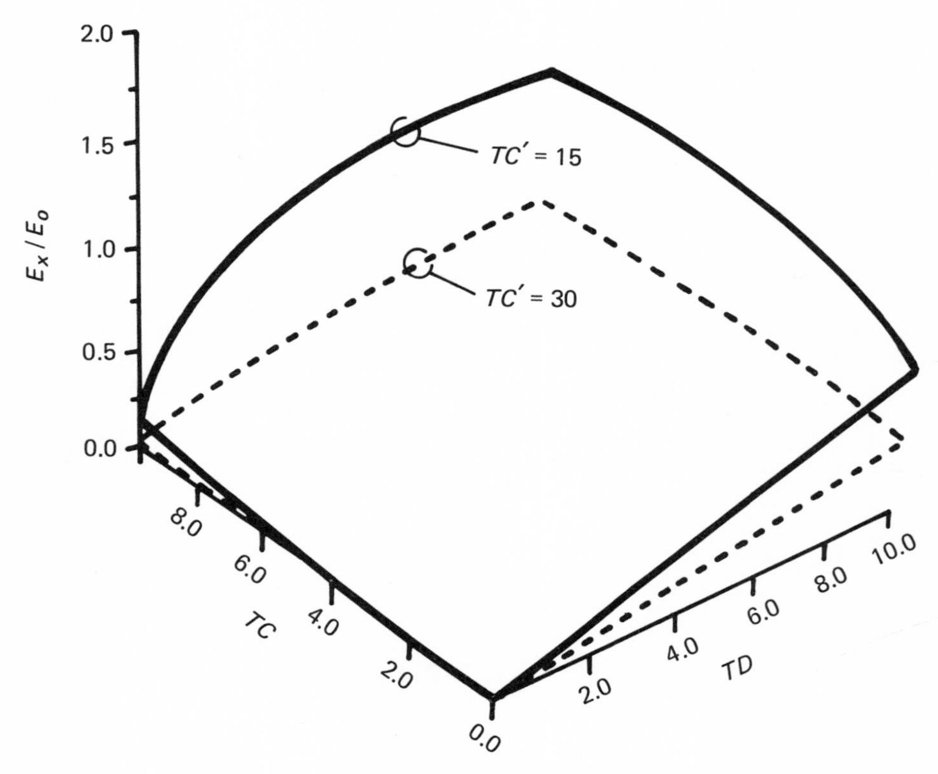


FIGURE 4.12
Three-dimensional peak error for a runaway process.

$K_v$ = valve gain
$K_p$ = process gain
$K_m$ = measurement gain

## 4.6
## *Examples*

### 4.6.1 WASTE TREATMENT pH LOOP (SELF-REGULATING)

*Given:*

(a) Set point is 7 pH.
(b) Measurement range is 0 to 14 pH.
(c) Minimum influent flow is 10 gallons per minute (gpm) (*Fil*).
(d) Normal influent flow is 22 gpm (*Fin*).
(e) Maximum influent flow is 100 gpm (*Fih*).
(f) Influent concentration is 32 percent by weight HCl—10.17 normality ($C_i$).
(g) Influent disturbance is rapid 20-gpm increase in flow ($\Delta L$).
(h) Reagent concentration is 20 percent by weight NaOH—7.93 normality ($C_r$).
(i) Vertical tank liquid volume is 1000 gallons ($V$).
(j) Axial blade agitator diameter is 1 foot ($D$).
(k) Axial blade agitator speed is 120 rpm ($N_s$).
(l) Axial blade agitator discharge coefficient is 1 ($N_q$).

*Find:* peak and accumulated errors for one, two, and three tanks in series with individual control loops and with just one overall control loop

*Solution:*

(a) Calculate the ultimate period for the individual loops:

$$F_a = 7.48*N_q*N_s*D^3$$

$$F_a = 7.48*1*120*1^3 = 898 \text{ gpm} \quad \text{(E4)}$$

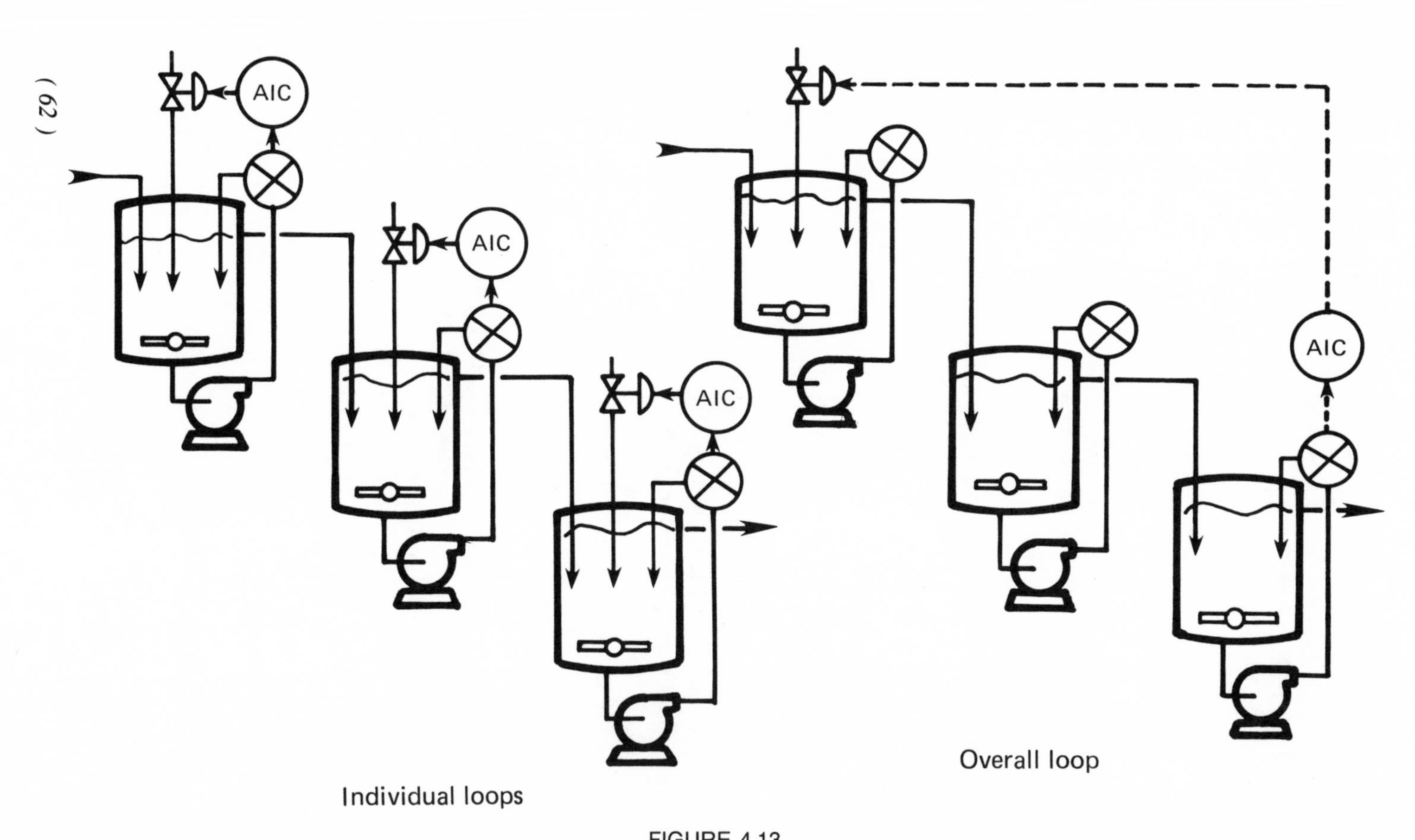


FIGURE 4.13
Waste treatment pH loop (self-regulating).

($F_a$ is the equivalent flow due to agitation—see Appendix E.)

$$F_{rl} = \frac{C_i}{C_r}*F_{il} = \frac{10.17}{7.93}*10 = 12.28 \text{ gpm minimum reagent flow}$$

$$F_{rn} = \frac{C_i}{C_r}*F_{in} = \frac{10.17}{7.93}*22 = 28.21 \text{ gpm normal reagent flow}$$

$$F_{rh} = \frac{C_i}{C_r}*F_{ih} = \frac{10.17}{7.93}*100 = 128.25 \text{ gpm maximum reagent flow}$$

(The above equations are a normality balance for the neutralization of a strong acid by a strong base.)

$$F_{tn} = F_{rn} + F_{in} = 28.21 + 22$$
$$= 50.2 \text{ gpm total normal throughput flow}$$

$$TD = \frac{V}{F_{tn} + F_a} = \frac{1000}{50 + 898} \tag{E1}$$
$$= 1 \text{ minute for each tank}$$

(The above equation estimates the dead time due to imperfect mixing.)

$$TC = \frac{V}{F_{tn}} - TD = \frac{1000}{50} - 1$$
$$= 19 \text{ minutes for each tank} \tag{E2}$$

(The time constant is the residence time minus the mixing dead time. $V/F$ is the tank residence time—see Appendix E.)

$$T_u = 2*\left[1 + \left[\frac{TC}{TC + TD}\right]^{0.65}\right]*TD \tag{4.5}$$

$$T_u = 2*\left[1 + \left[\frac{19}{19 + 1}\right]^{0.65}\right]*1$$
$$= 4 \text{ minutes}$$

(b) Calculate the ultimate period for the overall loop:

$$TD = Y{*}N{*}TC$$

$$TD = 0.28{*}3{*}19 = 16 \text{ minutes for three tanks in series}$$

$$TC = \frac{N{*}TC - TD}{(1 + Y)} = \frac{3{*}19 - 16}{(1 + 0.28)} \tag{4.4}$$

$$= 32 \text{ minutes for three tanks in series}$$

$$T_u = 2{*}\left[1 + \left[\frac{TC}{TC + TD}\right]^{0.65}\right]{*}TD$$

$$T_u = 2{*}\left[1 + \left[\frac{32}{32 + 16}\right]^{0.65}\right]{*}16 \tag{4.5}$$

$$= 57 \text{ minutes}$$

(c) Calculate the open-loop steady-state error:

$$E_o = K_p{*}K_l{*}\Delta L \tag{3.2}$$

$$E_o = 1{*}10.17{*}0.20 = 2 \text{ normality}$$

(d) Calculate the peak and accumulated errors for the individual loops:

$$E_x = \left[\frac{1.1{*}K_g{*}T_u}{2\pi{*}TC}\right]{*}E_o$$

$$E_x = \left[\frac{1.1{*}1.5{*}4}{2\pi{*}19}\right]{*}2$$

$$E_x = 0.05{*}2 = 0.1 \text{ normality for first tank}$$

$$E_x = 0.05{*}0.1 = 0.005 \text{ normality for second tank}$$

$$E_x = 0.05{*}0.005 = 0.00025 \text{ normality for third tank}$$

$$E_x = 7 - (\text{pH for } 0.1 \text{ normality}) \tag{4.14}$$

$$= 7 - 1 = 6 \text{ pH for first tank}$$

$$E_x = 7 - (\text{pH for } 0.005 \text{ normality})$$

$$= 7 - 2.3 = 4.7 \text{ pH for second tank}$$

$$E_x = 7 - (\text{pH for } 0.00025 \text{ normality})$$

$$= 7 - 3.6 = 3.4 \text{ pH for third tank}$$

[The pH for a given normality can be calculated by iterative solution of equations in Shinskey (1973) or from a titration curve.]

$$E_i = \frac{E_x}{1.1}*0.5*T_u \quad \text{(Equation 4.14 substituted into Equation 4.13)}$$

$$E_i = \frac{6}{1.1}*0.5*4 = 11 \text{ pH*minutes for first tank}$$

$$E_i = \frac{4.7}{1.1}*0.5*4 = 8.5 \text{ pH*minutes for second tank}$$

$$E_i = \frac{3.4}{1.1}*0.5*4 = 6.2 \text{ pH*minutes for third tank}$$

(e) Calculate the peak and accumulated errors for the overall loop:

$$E_x = \frac{1.1*K_g*T_u}{2\pi*TC}*E_o$$

$$E_x = \frac{1.1*1.5*57}{2\pi*32}*2$$

$$E_x = 0.47*2 = 0.94 \text{ normality}$$

$$E_x = 7 - (\text{pH for 0.94 normality}) \quad (4.14)$$

$$= 7 - 0.02 = 6.98 \text{ pH}$$

$$E_i = \frac{E_x}{1.1}*0.5*T_u$$

$$E_i = \frac{6.98}{1.1}*0.5*57 = 181 \text{ pH*minutes}$$

*Conclusions:* The peak error for the overall loop (0.94 normality) is over 3700 times larger than the peak error for the individual loops (0.00025 normality) and the accumulated error is (57/4)*3000, or nearly 43,000 times larger in normality units. The increase in errors expressed in pH units does not seem as bad because of the exponential relationship between normality and pH. However, normality errors represent the reagent (process) errors, which are a better index of the loop's performance than the pH (measurement) errors. The use of pH or ORP for composition control increases the rangeability and sensitivity for set points near neutrality (dilute set points) but decreases the awareness of the loop's performance. Also, the nonlinearity of the measurement necessitates increasing the *PB* to match the highest measurement gain (at neutrality for pH loops), which increases the errors over the rest of the pH measurement range. For the above strong acid and strong base, the required *PB* at pH = 7 is much greater than the *PB* available. The setting of a notch gain in a nonlinear controller is discussed for this same example in Chapter 9.

### 4.6.2 BOILER FEEDWATER FLOW LOOP (SELF-REGULATING)

*Given:*

(a) Set point is 100,000 pounds per hour (pph).
(b) Measurement range is 0 to 200,000 pph.
(c) Disturbance is rapid 20 percent increase in flow ($\Delta L$).
(d) Pipe diameter is 4 inches or 0.33 foot ($D_p$).
(e) Fluid velocity is 5 feet per second (fps) ($V_f$).
(f) Pipe friction factor is 0.01 ($C_p$).
(g) Pipe wall modulus of elasticity is 500,000,000 lb/ft$^2$ ($E_p$).
(h) Pipe wall thickness is 0.34 inch or 0.03 foot ($H$).
(i) Fluid density is 62 lb/ft$^3$ ($W$).
(j) Fluid modulus of elasticity is 5,000,000 lb/ft$^2$ ($E_f$).
(k) Acceleration due to gravity is 32 ft/(sec*sec) ($G$).

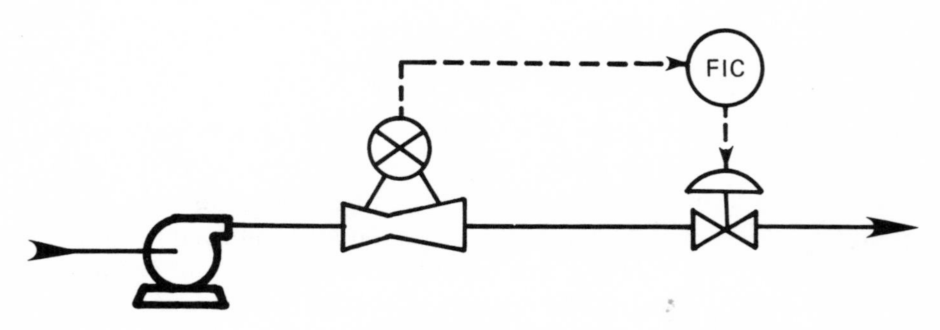


FIGURE 4.14
Boiler feedwater flow loop (self-regulating).

(l) Pipe length from valve to transmitter or discharge is 150 feet ($S_p$).
(m) Ratio of valve pressure drop to pipe pressure drop is 0.7 ($R$).

*Find:* peak and accumulated errors.
*Solution:*
(a) calculate the ultimate period for the loop:

$$TD = \frac{S_p}{V_w}$$

$$V_w = \left[ \frac{G}{W^* \left[ \frac{D_p}{E_p{}^*H} + \frac{1}{E_f} \right]} \right]^{0.5} = 1500 \text{ fps}$$

[$V_w$ is the wave velocity of pressure disturbances per equation in Bergeron (1961). Fluid does not start to accelerate at the transmitter until the pressure change due to the stroke change propagates from the valve to the transmitter by a plane wave.]

$$TD = \frac{150}{1500} = 0.1 \text{ second}$$

(This process dead time is less than the valve and transmitter dead time. Extremely large process dead times have been observed for fluids with high viscosity that flow through a small orifice such as a d/p integral orifice or a static mixer injection orifice.)

$$TC = \frac{D_p}{4^*C_p{}^*V_f{}^*(1 + R)}$$

$$TC = \frac{0.33}{4^*0.01^*5^*(1 + 0.7)} = 1 \text{ second}$$

[Fluid acceleration $TC$ is per equation in Harriot (1964). This equation is only applicable to turbulent flow.]

$$T_u = 2^*\left[1 + \left[\frac{TC}{TC + TD}\right]^{0.65}\right]^*TD$$

$$T_u = 2^*\left[1 + \left[\frac{1}{1 + 0.1}\right]^{0.65}\right]^*0.1 \qquad (4.5)$$

$$= 0.4 \text{ second}$$

(b) Calculate the open-loop steady-state error:

$$E_o = K_p{}^*K_l{}^*\Delta L$$

$$E_o = 1^*200000^*0.2 = 40000 \text{ pph} \qquad (3.2)$$

$$= 10 \text{ 1b/sec}$$

(c) Calculate the peak and accumulated errors:

$$E_x = \left[\frac{1.1^*K_g{}^*T_u}{2\pi^*TC}\right]^*E_o$$

$$E_x = \left[\frac{1.1^*1.5^*0.4}{2\pi^*1}\right]^*10 \qquad (4.14)$$

$$= 1 \text{ lb/sec} = 4000 \text{ pph}$$

$$E_i = \left[\frac{E_x}{1.1}\right]^*T_i$$ (Equation 3.11 substituted into Equation 3.7)

$$T_i = 0.8^*T_u = 0.8^*0.4 = 0.3 \text{ second} \qquad (4.7)$$

[Most controllers do not have a reset setting in excess of 100 repeats per minute or 0.6 second per repeat ($T_i$ = 0.6).]

$$E_i = \left[\frac{1}{1.1}\right]^*0.6 = 0.5 \text{ pound}$$

*Conclusions:* The accumulated and peak errors predicted above are much smaller than those experienced in the field because noise increases the usable *PB* and the dead time from the valve and transmitter dynamics increases $T_u$ significantly. Noise and instrument dynamics limit the flow loop's performance rather than process dynamics (valve hysteresis and characteristics and transmitter accuracy and nonlinearity may also limit the loop's performance; see Chapters 6 and 7).

### 4.6.3 BOILER DRUM LEVEL LOOP (INTEGRATING)

*Given:*
(a) Set point is 2 feet.
(b) Measurement range is 0 to 4 feet.
(c) Boiler feedwater flow dynamics are as described in Section 4.4.2.
(d) Level controller output goes directly to feedwater valve.
(e) Disturbance is rapid 20 percent decrease in steam flow.
(f) Boiler drum diameter is 4 feet ($D_d$).
(g) Boiler drum length is 10 feet ($S_d$).

*Find:* peak and accumulated errors.
*Solution:*
(a) Calculate the ultimate period for the loop.
From Example 4.4.2:

$$TD = \frac{S}{V_w} = 0.1 \text{ second}$$

$$TC = \frac{D_p}{4^*C_p{}^*V_f(1 + R)} = 1 \text{ second}$$

$$T_u = 4^*\left[1 + \left[\frac{TC}{TD}\right]^{0.65}\right]^*TD \tag{4.15}$$

$$T_u = 4^*\left[1 + \left[\frac{1}{0.1}\right]^{0.65}\right]^*0.1$$

$$= 2.2 \text{ seconds}$$

(b) Calculate the open-loop steady-state error:

$$K_i = \frac{1}{A^*W}$$

$$K_i = \frac{1}{D_d{}^*S_d{}^*W} = \frac{1}{4^*10^*62} \tag{C10}$$

$$= 0.0004 \text{ ft/lb at set point}$$

(The integrator gain will vary with level since the drum is horizontal; see Appendix C for the origin of the equation for $K_i$).

$$
\begin{aligned}
K_p &= 1*K_i = 0.0004 \text{ ft/lb at set point} \\
E_o &= K_p*K_l*\Delta L \qquad (3.2) \\
E_o &= 0.0004*200{,}000*0.2 \\
&= 16 \text{ ft/hr} = 0.0045 \text{ fps}
\end{aligned}
$$

[Level systems without inverse response, excessive noise, excessive dead time or time constant, and excessive valve hysteresis can be controlled with readily available hardware if the open loop error is less than 0.375 fps (Anderson, 1979).]

(c) Calculate the peak and accumulated errors.

Check the *PB*:

$$
\mathrm{K}_v = \frac{200000 \text{ lb/hr}}{100\%} = \frac{56 \text{ lb/sec}}{100\%} = \frac{0.56 \text{ lb/sec}}{\%}
$$

$$
\mathrm{K}_m = \frac{100\%}{4 \text{ ft}} = \frac{25\%}{\text{ft}}
$$

$$
PB = \frac{K_g*100*T_u^2*K_v*\mathrm{K}_p*\mathrm{K}_m}{(2\pi)^2*TC}
$$

$$
PB = \frac{1.5*100*2.2^2*0.56*0.0004*25}{(2\pi)^2*1} \qquad (4.16)
$$

$$
= 0.1 \text{ percent}
$$

(The minimum *PB* on many controllers is about 10 percent, which

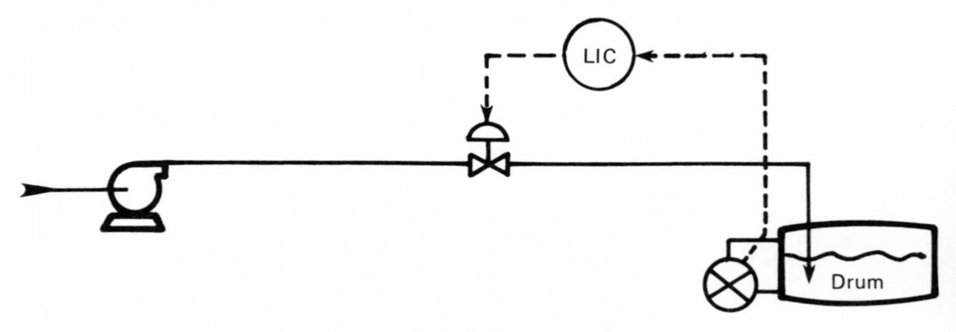


FIGURE 4.15
Boiler drum level loop (integrating).

is about 100 times larger than calculated. Chapter 4 stated that the period of oscillation is about 25 times larger than the original ultimate period $T_u$, the oscillations are nearly undamped, and the peak error is about ten times larger for this condition.)

$$E_x = 10*\left[\frac{1.1*K_g*T_u^2}{(2\pi)^2*TC}\right]*E_o$$

(10*Equation 4.20)

$$E_x = 10*\left[\frac{1.1*1.5*2.2^2}{(2\pi)^2*1}\right]*0.27$$

$$= 0.55 \text{ foot}$$

$$E_i = \left[\frac{E_x}{1.1}\right]*0.5*T_u \text{ (Equation 4.20 substituted into 4.19)}$$

$$E_i = \left[\frac{0.55}{1.1}\right]*0.5*(25*2.2)$$

$$= 13.75 \text{ feet*second}$$

*Conclusions:* Even though the controller *PB* lower limit decreased the loop performance, the accumulated and peak errors predicted above are still smaller than those experienced in the field because noise and inverse response increase the usable *PB* and the dead time from the valve and transmitter dynamics increases the $T_u$ significantly. Noise, inverse response, and instrument dynamics limit the level loop's performance rather than process dead time (valve hysteresis can cause a limit cycle in an integrating process; see Chapter 7).

### 4.6.4 FURNACE PRESSURE LOOP (PSEUDOINTEGRATING)

*Given:*

(a) Set point is 5 inches water column (w.c.) gage.
(b) Measurement range is 0 to 10 inches w.c. gage.
(c) Furnace volume is 10,000 ft$^3$ ($V_f$).
(d) Quench volume is 1000 ft$^3$ ($V_q$).
(e) Scrubber volume is 1000 ft$^3$ ($V_s$).

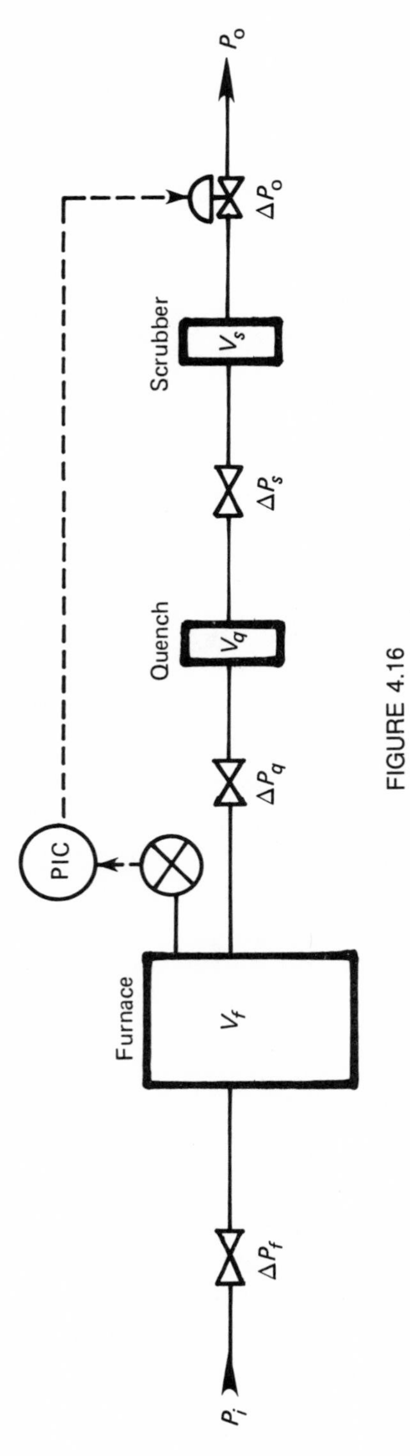


FIGURE 4.16
Furnace pressure loop (pseudointegrating).

(f) Furnace inlet flow resistance pressure drop is 2.5 inches w.c. ($\Delta P_f$).
(g) Quench inlet flow resistance pressure drop is 5 inches w.c. ($\Delta P_q$).
(h) Scrubber inlet flow resistance pressure drop is 10 inches w.c. ($\Delta P_s$).
(i) System outlet flow resistance pressure drop is 2.5 inches w.c. ($\Delta P_o$).
(j) Flue gas flow is 1000 scfm ($F_f$).
(k) Atmospheric pressure is 408 inches w.c. ($P_a$).
(l) Disturbance is a rapid 20 percent increase in inlet pressure ($\Delta L$).
(m) Inlet pressure (discharge of forced draft fan) is 15 inches w.c. ($P_i$).

*Find:* peak and accumulated errors.
*Solution:*
(a) Calculate the ultimate period:

$$R_f = \frac{2^*\Delta P_f}{F_f} = \frac{5}{1000} = 0.005 \text{ inch w.c. per scfm}$$
$$R_q = \frac{2^*\Delta P_q}{F_f} = \frac{10}{1000} = 0.01 \text{ inch w.c. per scfm}$$
$$R_s = \frac{2^*\Delta P_s}{F_f} = \frac{20}{1000} = 0.02 \text{ inch w.c. per scfm}$$
$$R_o = \frac{2^*\Delta P_o}{F_f} = \frac{5}{1000} = 0.005 \text{ inch w.c. per scfm}$$
$$C_f = \frac{V_f}{P_a} = \frac{10{,}000}{408} = 24.5 \text{ scf per inch w.c.}$$
$$C_q = \frac{V_q}{P_a} = \frac{1000}{408} = 2.45 \text{ scf per inch w.c.}$$
$$C_s = \frac{V_s}{P_a} = \frac{1000}{408} = 2.45 \text{ scf per inch w.c.}$$

(See Appendix G for the equations for the resistances and capacitances.)

$$TC1 = \frac{2*A}{B + (B^2 - 4*A*C)^{0.5}} \tag{F1}$$

$$TC2 = \frac{2*A}{B - (B^2 - 4*A*C)^{0.5}} \tag{F2}$$

$$\begin{aligned} A &= R_q*C_q*R_s*C_s*R_o \\ A &= 0.01*2.45*0.02*2.45*0.005 \\ &= 0.000006 \end{aligned} \tag{F3}$$

$$\begin{aligned} B &= R_q*C_q*R_o + R_q*C_q*R_s + R_s*C_s*R_o \\ &\quad + R_q*C_s*R_o \\ B &= 0.01*2.45*0.005 + 0.01*2.45*0.02 \\ &\quad + 0.02*2.45*0.005 \\ &\quad + 0.01*2.45*0.005 \\ B &= 0.0001225 + 0.00049 + 0.000245 \\ &\quad + 0.0001225 = 0.001 \end{aligned} \tag{F4}$$

$$C = R_q + R_s + R_o \tag{F5}$$

$$C = 0.01 + 0.02 + 0.005 = 0.035$$

$$TC1 = \frac{2*0.000006}{0.001 + 0.0004} = 0.0086 \text{ minute}$$

$$= 0.5 \text{ second}$$

$$TC2 = \frac{2*0.000006}{0.001 - 0.0004} = 0.02 \text{ minute}$$

$$= 1.2 \text{ seconds}$$

(See Appendix F for the equations for the noninteractive time constants.)

$$\begin{aligned} TC &= TC2 = 1.2 \text{ seconds} \\ TD &= Y*TC1 \\ TD &= 0.4*0.5 = 0.2 \text{ second} \end{aligned} \tag{4.2}$$

$$T_u = 4*\left[1 + \left[\frac{TC}{TD}\right]^{0.65}\right]*TD$$

$$T_u = 4*\left[1 + \left[\frac{1.2}{0.2}\right]^{0.65}\right]*0.2 \qquad (4.15)$$

$$= 3.4 \text{ seconds}$$

(b) Calculate the open-loop steady-state error:

$$K_i = \frac{K_f}{TC_f} = \frac{R_q}{R_f + R_q} * \frac{R_f + R_q}{R_q * R_f * C_f}$$

$$= \frac{1}{(R_f * C_f)}$$

$$K_i = \frac{1}{(R_f * C_f)} = \frac{1}{(0.005*24.5)}$$

$$= 8.2 \text{ per minute}$$

(See Appendix C for the equation for the pseudointegrator gain.)

$$K_p = 1*8.2 = 8.2 \text{ per minute}$$

$$E_o = K_p * K_l * \Delta L \qquad (3.2)$$

$$E_o = 8.2*10*0.2 = 16.3 \text{ inches w.c. per minute}$$

$$E_o = 0.29 \text{ inch w.c. per second}$$

(c) Calculate the peak and accumulated errors:

$$E_x = \left[\frac{1.1*K_g*T_u^2}{(2\pi)^2*TC}\right]*E_o$$

$$E_x = \left[\frac{1.1*1.5*3.4^2}{(2\pi)^2*1.2}\right]*0.29$$

$$= 0.12 \text{ inch w.c.} \qquad (4.20)$$

$$E_i = \left[\frac{E_x}{1.1}\right]*0.5*T_u$$ (Equation 4.20 substituted into 4.19)

$$E_i = \left[\frac{0.12}{1.1}\right]*0.5*3.4 = 0.18 \text{ (inch w.c.)*second}$$

*Conclusions:* The accumulated and peak errors are small because the integrator gain is small since the furnace volume is large and the outlet volumetric flow will increase as the pressure in the system increases. If the volumetric outlet flow were fixed by an induced draft fan, the integrator gain would increase tremendously. The closed loop errors are smaller than those experienced in the field because measurement noise, measurement dynamics, control valve dynamics, and interaction limit the performance rather than process dead time (see Chapters 7 and 8).

### 4.6.5 EXOTHERMIC REACTOR TEMPERATURE LOOP (RUNAWAY)

*Given:*

(a) Set point is 150 F.
(b) Measurement range is 100 to 200 F.
(c) Reactant feed flow is 2000 pph ($W_r$).
(d) Reactant mass is 3000 pounds ($M_r$).
(e) Reactant heat capacity is 0.5 Btu/(lb*F) ($C_r$).

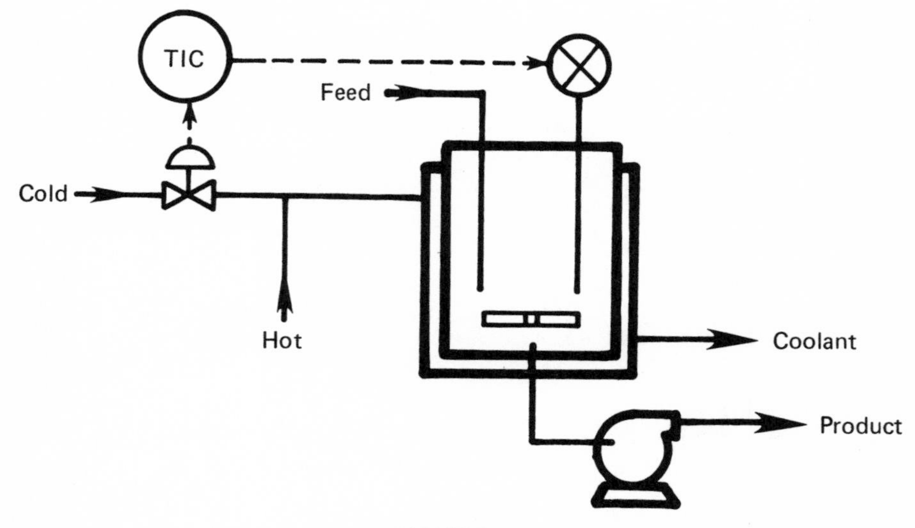


FIGURE 4.17
Reactor temperature loop (runaway).

(f) Heat transfer coefficient*area is 8000 Btu/(hr*F) ($UA$).
(g) Change in heat generation with temperature is 12,000 Btu/F ($\Delta Q/\Delta T$).
(h) Coolant mass is 400 pounds ($M_c$).
(i) Coolant heat capacity is 1 Btu/(lb*F) ($C_c$).
(j) Coolant flow is 80,000 pph ($W_c$).
(k) Disturbance is a 20 percent increase in coolant temperature ($\Delta L$).
(l) Coolant temperature is 100 F.
(m) Continuous reactor.

*Find:* peak and accumulated errors.
*Solution:*
(a) Calculate the ultimate period for the loop:

$$T_u = 4*\left[1 + \left[\frac{N}{D}\right]^{0.65}\right]*TD$$

$$N = (TC' + TC)*TC'*TC \qquad (4.21)$$

$$TC' = \frac{M_r*C_r}{UA + W_r*C_r - \dfrac{\Delta Q}{\Delta T}}$$

(See Table 4.1 for source of equation.)

$$TC' = -\frac{3000*0.5}{8000 + 2000*0.5 - 12{,}000}$$

$$= 0.5 \text{ hour} = 30 \text{ minutes}$$

$$TC = \frac{M_c*C_c}{UA}$$

(See Table 4.1 for source of equation.)

$$TC = \frac{400*1}{8000} = 0.05 \text{ hour} = 3 \text{ minutes}$$

$$TD = \frac{M_c}{W_c} = \frac{400}{80{,}000} = 0.005 \text{ hour}$$

$$= 0.3 \text{ minute}$$

$$N = (30 + 3)*30*3 = 2970$$
$$D = (TC' - TC)*(TC' - TD)*TD$$
$$D = (30 - 3)*(30 - 0.3)*0.3$$
$$= 241$$
$$T_u = 4*\left[1 + \frac{2970}{241}^{0.65}\right]*0.3$$
$$= 7.4 \text{ minutes}$$

(b) Calculate the open-loop steady-state error:

$$E_o = K_p*K_l*(\Delta L)$$
$$K_p = -\frac{UA}{UA + W*C_r - \dfrac{\Delta Q}{\Delta T}}$$
$$K_p = -\frac{8000}{8000 + 2000*0.5 - 12{,}000} \quad (3.2)$$
$$= 2.7$$
$$E_o = 2.7*100*(0.2) = 53 \text{ F}$$

(c) Calculate the peak and accumulated errors:

$$E_x = \left[\frac{1.1*K_g*T_u^2}{(2\pi)^2*TC*TC'}\right]*E_o$$
$$E_x = \left[\frac{1.1*1.5*7.4^2}{(2\pi)^2*3*30}\right]*53$$
$$= 1.4 \text{ F} \quad (4.25)$$
$$E_i = \left[\frac{E_x}{1.1}\right]*0.5*T_u$$ (Equation 4.20 substituted into 4.19)
$$E_i = \left[\frac{1.4}{1.1}\right]*0.5*7.4 = 4.7 \text{ deg/min}$$

*Conclusions:* The accumulated and peak errors predicted above are smaller than those experienced in the field because the time constant of the thermowell and temperature bulb increases the dead time $TD$ and hence the ultimate period $T_u$ (see Chapter 6). Also, the heat transfer coefficient tends to decrease with time, which increases $TC$ and $TC'$ and hence increases the ultimate period $T_u$.

### 4.6.6 BIOLOGICAL REACTOR CONCENTRATION LOOP (RUNAWAY)

*Given:*

(a) Average nutrient concentration is 4 nanograms per milliliter (ng/ml) ($C_i$).
(b) Nutrient concentration where growth rate is half its maximum is 1 ng/ml ($K_i$).
(c) Maximum cell growth rate is 0.01 generation per minute ($U_x$).
(d) Substrate consumption rate for cell survival is 0.001 generation per minute ($K_e$).
(e) Reactor working volume is 600 gallons ($V$).
(f) Average throughput flow is 10 gpm ($F$).

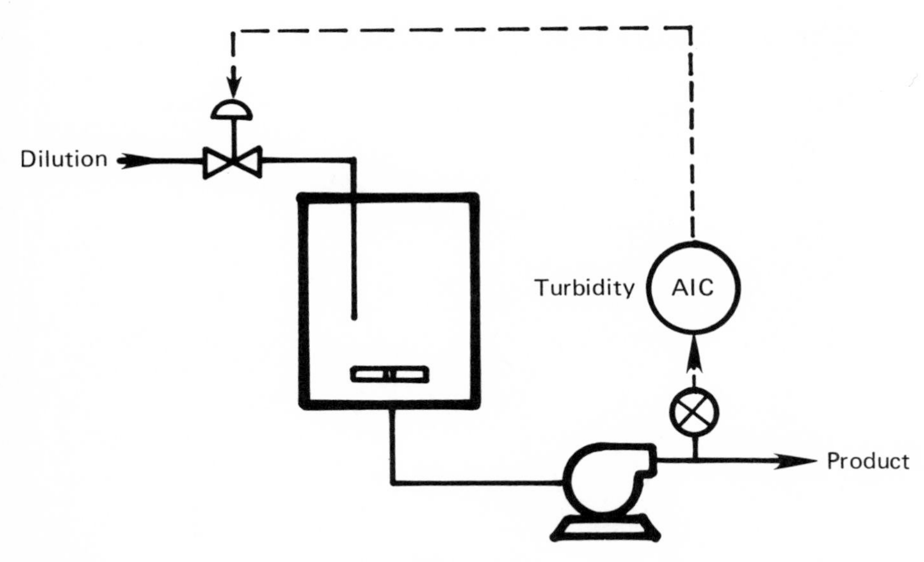


FIGURE 4.18
Biological reactor concentration loop (runaway).

(g) Agitator pumping rate is 200 gpm ($F_a$).
(h) Loop steady-state gain is 1 ($K_v$*$K_p$*$K_m$).

*Find:* the proportional band window size
*Solution:*
(a) Calculate the ultimate period for the loop:

$$T_u = 4*\left[1 + \left[\frac{N}{D}\right]^{0.65}\right]*TD$$

$$N = (TC' + TC)*TC'*TC$$

$$D = (TC' - TC)*(TC' - TD)*TD$$

$$TD = \frac{V}{F_a + F} = \frac{600}{200 + 10} = 3 \text{ minutes} \qquad (4.21)$$

$$TC = \frac{V}{F} - TD = \frac{600}{10} - 3 = 57 \text{ minutes}$$

$$TC' = 1/U$$

$$U = U_x*\left[\frac{C_i}{K_i + C_i}\right] - K_e = 0.01*\left[\frac{4}{1 + 4}\right] - 0.001 = 0.007$$

(See Table 4.1 for source of the equation.)

$$TC' = 1/0.007 = 142 \text{ minutes}$$

$$N = (142 + 57)*142*57 = 1{,}610{,}706$$

$$D = (142 - 57)*(142 - 3)*3 = 35{,}445$$

$$T_u = 4*\left[1 + \left[\frac{1{,}610{,}706}{35{,}445}\right]^{0.65}\right]*3 = 155 \text{ minutes}$$

(b) Calculate the minimum and maximum proportional bands:

$$PB_{\text{max}} = 80*K_v*K_p*K_m = 80 \text{ percent} \qquad (4.26)$$

$$PB_{\text{min}} = 0.5*PB \qquad (4.27)$$

$$PB = \frac{K_g*100*T_u^2*K_v*K_p*K_m}{(2\pi)^2*TC*TC'} = \frac{3{,}603{,}750}{319{,}214} = 11 \text{ percent} \qquad (4.22)$$

$$PB_{\text{min}} = 0.5*11 = 5.5 \text{ percent}$$

(c) Calculate the *PB* window size:

$$\frac{PB_{max}}{PB_{min}} = \frac{80}{5.5} = 14.5$$

*Conclusions:* The proportional band window size is slightly less than the minimum window size of 15 recommended by Luyben (Luyben, 1978). An increase in nutrient concentration will increase the growth rate, which will decrease the runaway time constant, which will decrease the window size. However the growth rate is already close to its maximum. Any analysis measurement dead time for nutrient dilution control or substrate activation delays will decrease the window size. The ultimate period is so large that tuning will be tedious. Also, the integral and derivative time settings required are beyond the range of most conventional analog controllers. If the integral time setting used is too small, the loop will go into a slow reset cycle.

CHAPTER V

# *Effect of Controller Dynamics*

## *5.1 Parallel Controller Algorithm*

The parallel controller computes the proportional, integral, and derivative modes in parallel (see Figure 5.1). Thus the modes are noninteractive in the time domain but interactive in the frequency domain. The parallel controller is referred to as the "ideal" or "noninteractive" controller by Shinskey and others. The accumulated error for a parallel controller can be about one half that for a series controller. However, if the derivative and integral times are set close together, the controller becomes extremely sensitive to the inevitable changes in the loop gain of chemical processes and the loop period will drift (the accumulated error will also drift). The Foxboro Company tested a parallel controller on a self-regulating process and found that the loop period increased from 16 to 55 seconds when the *PB* was changed from 10 percent to 100 percent with no change in the quarter-amplitude damping (Shinskey, 1979).

The derivative time can also be set larger than one fourth the integral time in a parallel controller, which creates zeroes

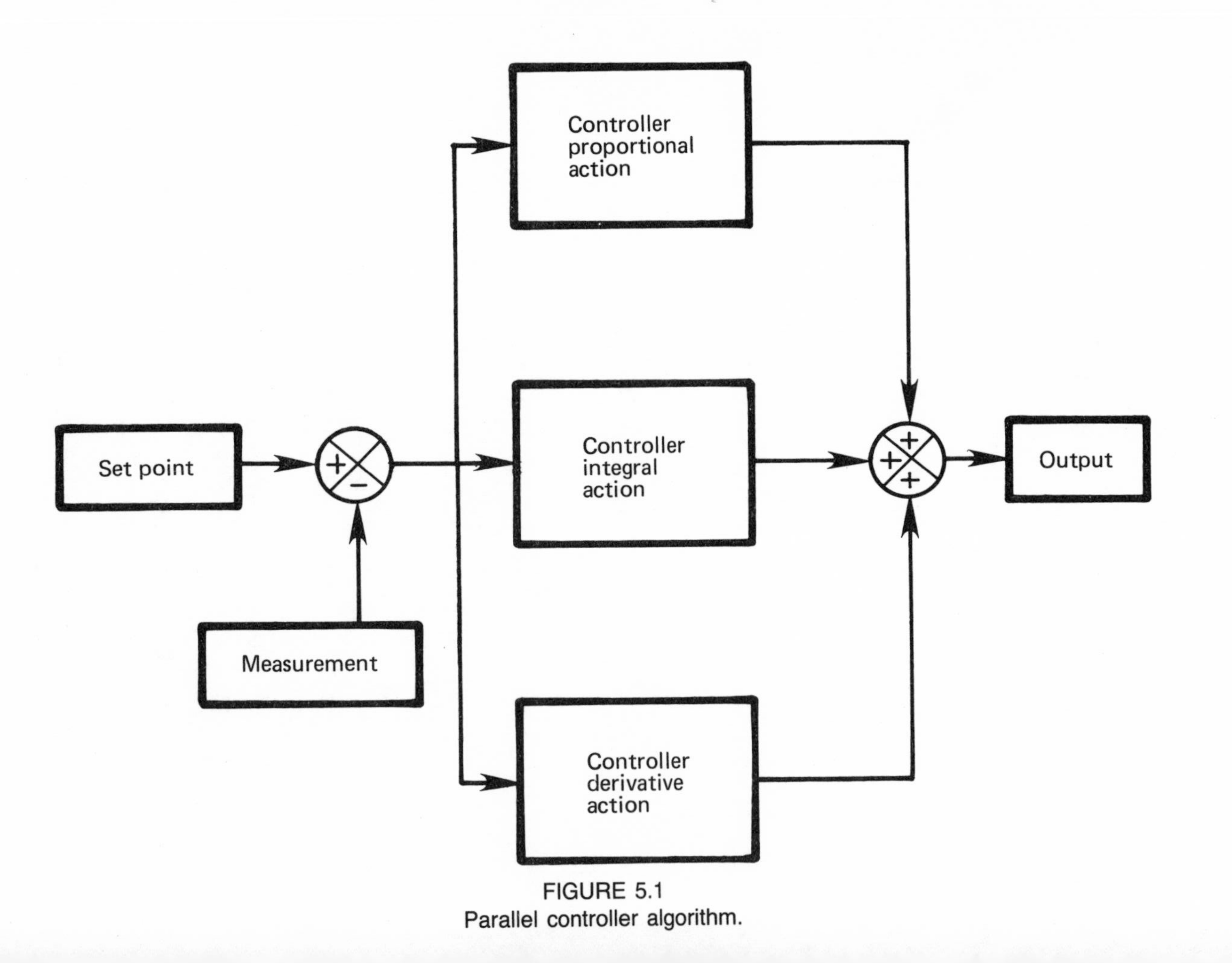


FIGURE 5.1
Parallel controller algorithm.

with imaginary parts in the controller's transfer function. Parallel controllers are difficult to tune and keep tuned and are seldom used. The equation for the parallel controller is frequently used in lectures and textbooks to explain controller mode actions because the mode actions are separable and simply represented. The equations for the peak and accumulated error presented previously use the mode settings of the parallel controller. This chapter shows that when the mode settings of a series controller are converted to the equivalent mode settings of a parallel controller and substituted into the equation for the accumulated error, the mode interaction factor cancels out (see Equations 5.1 and 5.2).

## 5.2 *Series Controller Algorithm*

The series controller computes the derivative mode in series with the integral and proportional modes (see Figure 5.2). Thus the modes are interactive in the time domain but noninteractive in the frequency domain. The series controller is referred to as the "real" or "interactive" controller by Shinskey and others. Some analog and many digital controllers have the derivative mode calculated before the integral mode to reduce the peak error. This feature is frequently referred to as "rate before reset." The equivalent parallel controller's derivative time, $T_d$, cannot be made larger than one fourth the equivalent parallel controller's integral time, $T_i$, because $T_i$ increases faster than $T_d$ when the series controller's integral time, $T_i'$, is increased. Since $T_d$ cannot be greater than 1/4 $T_i$, the zeroes of the controller's transfer function will be real. The equivalent parallel controller's mode settings can be calculated from the series controller's mode settings by use of the following equations.

$$PB = PB'*I_c \quad (5.1)$$

$$T_i = \frac{T_i'}{I_c} \quad (5.2)$$

$$T_d = T_d'*I_c \quad (5.3)$$

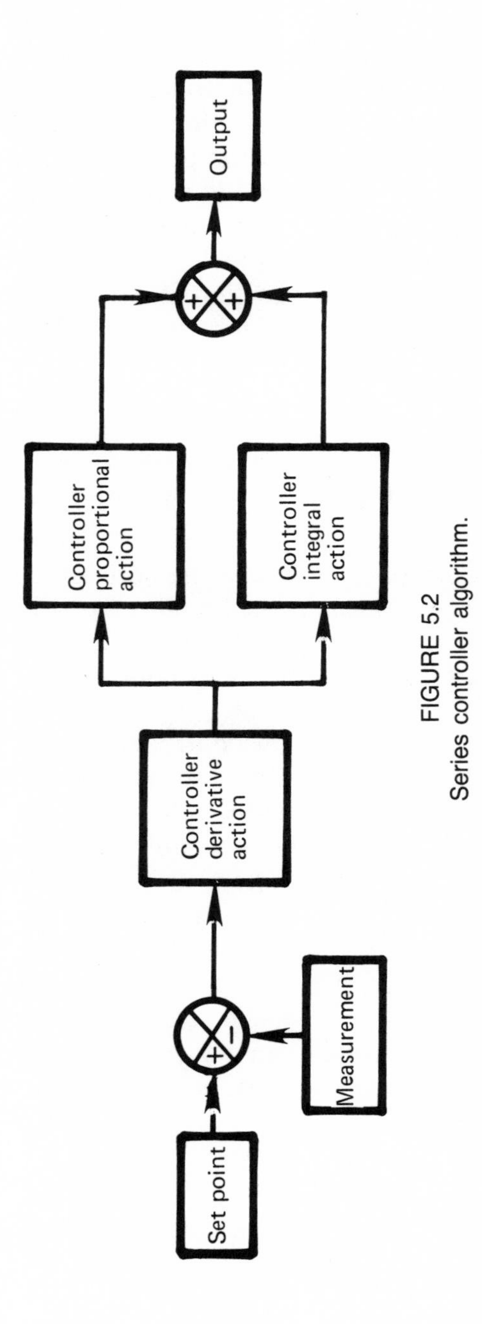


FIGURE 5.2
Series controller algorithm.

where

$$I_c = \frac{T'_i}{(T'_i + T'_d)}$$

$I_c$ = controller mode interaction factor
$PB$ = parallel controller's proportional band
$T_i$ = parallel controller's integral time
$T_d$ = parallel controller's derivative time
$PB'$ = series controller's proportional band
$T'_i$ = series controller's integral time
$T'_d$ = series controller's derivative time

The above equations show that for a three-mode (PID) controller, the equivalent parallel controller's proportional band and derivative time are smaller but the integral time is larger than the corresponding mode settings of the series controller. Foxboro recommends setting $T'_i$ equal to $T'_d$ on its controllers, which yields $T_d$ equal to 1/4 $T_i$ and $PB$ equal to 1/2 $PB'$ (Shinskey, 1979). Substitution of Equations 5.1 and 5.2 into Equation 3.1 for the accumulated error will result in the appearance of the controller mode interaction factor, $I_c$, in both the numerator and denominator.

Substitute Equations 5.1 and 5.2 into Equation 3.1 and simplify:

$$E_i = \frac{PB}{100*K_m}*T_i*\Delta C$$

$$E_i = \frac{PB'*I_c}{100*K_m}*\frac{T'_i}{I_c}*\Delta C \qquad (3.1)$$

$$= \frac{PB'}{100*K_m}*T'_i*\Delta C$$

Fortunately the controller mode interaction factor can be canceled out so that the equations previously developed for the accumulated error can be used for either parallel or series controllers. Most industrial analog and digital controllers are series controllers.

## 5.3 Analog Controllers

Analog controllers use continuous signals to compute the controller output. Most companies use electronic rather than pneumatic controllers on loops important enough to bring into the control room. Measurements by the author of the open loop

TABLE 5.1
Analog Controller Proportional Band Test Results

| *Dial PB, %* | *Measured PB, %* |
|---|---|
| 20 | 25 |
| 40 | 48 |
| 60 | 70 |
| 80 | 87 |
| 100 | 103 |
| 200 | 246 |
| 400 | 410 |
| 600 | 667 |
| 800 | 800 |
| 1000 | 1000 |

TABLE 5.2
Analog Controller Integral Time Test Results

| *Dial $T_i$, minute* | *Measured $T_i$, minute* |
|---|---|
| 0.2 | 0.5 |
| 0.1 | 0.2 |
| 0.05 | 0.1 |

TABLE 5.3
Analog Controller Derivative Time Test Results

| *Dial $T_d$, minute* | *Measured $T_d$, minute* |
|---|---|
| 0.05 | 0.03 |
| 0.1 | 0.03 |
| 0.2 | 0.08 |
| 0.5 | 0.12 |

time response of one industrial electronic analog controller revealed the following characteristics:

1. The measured proportional band was from 0 to 25 percent larger than the dial marking.

2. The measured integral time was about 100 percent larger than the dial marking.

3. The measured derivative time was from 40 to 70 percent smaller than the dial marking.

4. The measured integral time did not change with a change in the derivative dial setting. According to the equations presented in this chapter for series controller algorithms, the integral time should have increased as the derivative time increased. The measured derivative time and measured proportional band did approximately obey the equations except that the measured change was smaller than the calculated change for large dial settings.

5. The measured controller output showed a peak whenever a derivative setting of any value was used whereas simulation results of a series controller algorithm only showed a peak if

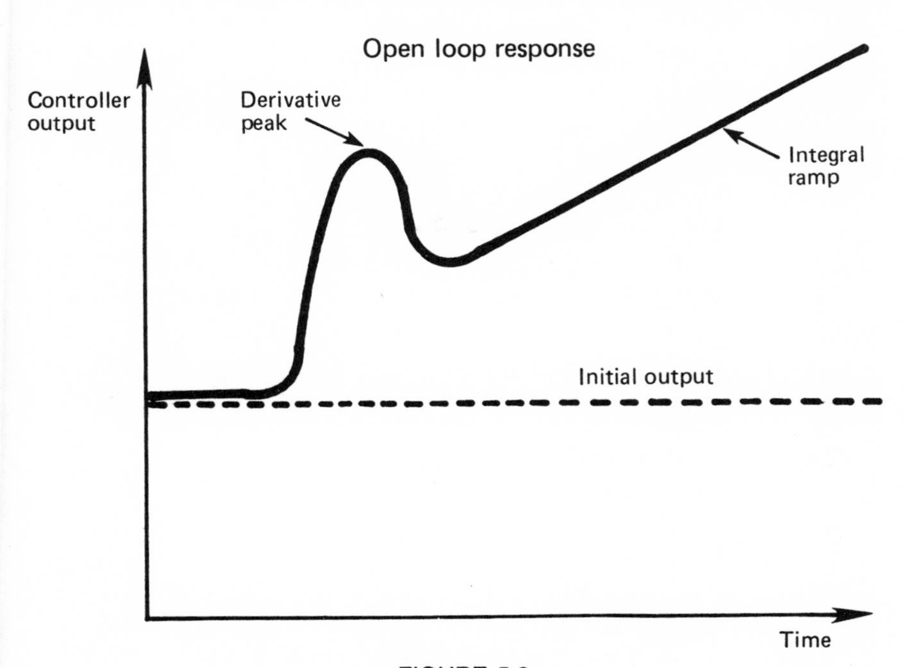


FIGURE 5.3
Analog controller open loop test results.

the derivative time was set larger than one fourth of the integral time (the disturbance was first order).

Not all electronic analog controllers show the above characteristics. However it is fair to say that the accuracy of the mode settings of analog controllers is not as good as that of digital controllers. Analog pneumatic controllers were used predominantly through 1970 and analog electronic controllers were used predominantly through 1980. Major projects designed after 1980 typically use digital controllers as part of a distributed control system.

## 5.4 *Digital Controllers*

Digital controllers use discrete (sample and hold) signals to compute the controller output. The microprocessor-based controllers in distributed control systems are digital controllers with algorithms that emulate the "real" three-mode analog controllers. They also have increased computational capability for feedforward control and signal or mode characterization. If the sample time, $T_s$, is less than one half of the ultimate period, $T_u$, then the effect of sampling is to add a dead time $TD$ to the loop that is equal to one half of the dead time. Extrapolator software has been proposed to reduce this effective dead time due to the conventional zero-order hold used in sampling. The extrapolator holds the sampler output constant at the expected value at the middle rather than the measured value at the beginning of the interval (Yekutiel, 1980).

For zero-order hold if $T_s < \frac{T_u}{2}$:

$$TD = \frac{T_s}{2} \tag{5.4}$$

There is also a computational dead time, a dead time due to the analog filter time constant used to filter out the harmonics of the sampling frequency, and a dead time due to the time constant used in the derivative mode computation used to reduce derivative noise (rate limiting). The summation of all of these dead times can be approximated as follows:

For digital controllers if $T_s < \frac{T_u}{2}$:

$$TD = \frac{T_s}{2} + \frac{T_s}{5} + Y^*TC + Y^*\frac{T_d}{a} \tag{5.5}$$

where

$TD$ = effective dead time added to the loop
$T_s$ = sample time of the digital controller
$TC$ = time constant of the analog sample filter
$T_d$ = derivative time setting
$a$ = rate-limiting factor for derivative time
$Y$ = the effective dead-time factor from Equation 4.1

There is also a digital software filter with an adjustable time constant that adds an equivalent amount of dead time. An adaptive delay digital filter can be designed that smooths digital signals with a bidirectional discriminator to reduce the filter time constant, and hence the equivalent dead time, by a factor of 16 for large signal changes (Loeber, 1980). This adaptive delay assumes that the signal to noise ratio is sufficient to discriminate. The effect of the additional dead time is negligible for an ultimate period of less than 4.0 seconds, if the sample time is 0.1 second or less. The minimum available derivative and integral time setting is usually proportional to the sample time. Except for very fast sample times, the minimum derivative and integral time settings of digital controllers are slightly larger than those of analog controllers. If a controller's derivative time is set equal to $0.125^*T_u$ per Equation 4.10, then the minimum ultimate period for three-mode control is about 6.0 seconds for a digital controller with a sample time of 0.1 second or more. The maximum integral time and derivative time are also proportional to the sample time. These maximums are usually larger in digital controllers than in analog controllers so that larger ultimate periods can be controlled.

Digital controllers should not be used indiscriminately on fast critical loops such as low-range furnace pressure and compressor surge control loops because of additional dead time. Although most flow loops are fast, the deterioration in loop performance resulting from the additional dead time and lower limits on mode settings of digital controllers is unimportant and is usually masked by appreciable measurement noise.

If the sample time, $T_s$, is greater than one half of the ultimate period, $T_u$, the dead-time approximation discussed above may not be accurate and aliasing can occur. Aliasing produces low-frequency components in the controller output signal, which correspond to the differences between the measurement input frequency and the sampling frequency or its harmonics (Goff, 1966). If a time constant greater than the ultimate period exists in the loop, the low-frequency components produced by aliasing are filtered so that they do not affect loop performance and the dead-time approximation discussed is again valid (Moore, 1969). However, if the time constant is upstream of the frequency source, which is typical for measurement noise, then the low-frequency components from aliasing can wear out the control valve or its accessories even though the process output may not be affected because of its large time constant. The loop gain can significantly amplify the amplitude of the alias. A loop with an integrating process, a controller gain of 18, a reset setting of ten repeats per minute, a sample frequency of 10 hertz (Hz), and a 0.5 percent noise signal of 9.75 Hz will produce a 17 percent alias of 0.25 Hz (the alias frequency is equal to the sample frequency minus the noise frequency). A Bessel measurement filter is better than a Butterworth measurement filter for attenuation of noise and aliasing because the dead time and time constant of the Bessel filter are less for a given attenuation. For example, a four-pole Bessel filter from Frequency Devices, Inc., will have a dead time of 0.17 second and a time constant of 0.22 second whereas a four-pole Butterworth will have a dead time of 0.22 second and a time constant of 0.30 second for 3-decibel attenuation at 1 Hz. Also, the open loop response of the four-pole Butterworth shows some ringing (open loop oscillation). Digital software filters cannot prevent aliasing. Aliasing can be prevented only by installing an analog filter on the signal input to the digital controller. Trend plot displays on the CRT of a distributed control system should not be used for the tuning of fast loops. The slow reporting rate can cause low-frequency aliases for the closed-loop ultimate-oscillation method or inaccurate dead-time estimates for the open-loop first-order-plus-dead-time method (Heider, 1982).

Digital controllers are superior to analog controllers in the accuracy and resolution of the mode settings. Digital controllers also are superior to analog controllers in the accuracy of auxiliary computations such as signal linearization and signal compensation (see Chapter 11 on signal characterization). As a result, the digital controller does not negate any extra effort made during design to improve loop performance (provided the sample time is fast enough). From a project cost standpoint, the flexibility in function and communication of digital controllers can save time and money in reduced wiring and panel costs for the initial design, and especially for any future modifications. However the increase in flexibility also results in an increase in responsibility for the engineer, since more choices mean more opportunity for mistakes.

ANTIRESET

## 5.5 *Antirest Windup Algorithms*

Reset windup occurs when a limit in the I/P or positioner output, valve stroke, process output, transmitter output, or controller input (typically imposed by hardware design, hardware calibration, interlock, or operating condition) prevents the controller from either reducing or seeing the error response so that the long-term error causes reset to drive the controller output to an extremity (although reset windup is usually associated with the upper extremity, it can also occur at the lower extremity). Reset action will not reverse the direction of the controller output until the measurement crosses the set point (i.e., the error changes sign). Proportional action will reverse the direction of the output when the controller input reverses, if the output does not exceed the output range.

If the controller output range is much larger than the range of the other loop components, the reset contribution may be considerably larger than the proportional contribution to the total controller output. If the output was driven upscale, saturation of amplifiers in electronic analog controllers can take place. Reset windup commonly occurs in override, surge, batch, and pH control loops. The duration of the windup is longer for those processes with large ultimate periods and large process gains because the small proportional and reset action prolongs the reversal of controller output. Most controllers have an anti-reset windup option that limits the contribution of reset action so that the summation of proportional plus reset action does not exceed a specified limit on either end of the output range (if the output limits are mistakenly set so that they overlap, the output will be frozen).

The addition of reset action to an analog electronic controller causes the proportional band to float with the measurement so that a change in controller input results in a change in the proportional contribution to the output, no matter how far away the measurement is from set point. For controllers with small proportional bands such as temperature controllers, the output may change full scale for a small change in measurement far away from set point. This rapid full-range change in output may be incorrectly attributed to derivative action or a hardware problem. The proportional band is not centered about the set point but is located just below the measurement. The following set of equations estimates the location of the proportional band with respect to the measurement for these controllers with and without the antireset option.

$$C = \frac{100}{PB} * \left[ (E - E_o) + \left( \int_{t_o}^{t} \frac{E}{T_i} * dt \right) \right] + C_o \tag{5.6}$$

For $C > (C_h + C_l)/2$ (direct acting):

$$X_h = \{PB * (C_h - C)\}/100 \tag{5.7}$$

$$X_l = X_h - PB \tag{5.8}$$

For $C < (C_h + C_l)/2$ (direct acting):

$$X_l = \{PB^*(C_l - C)\}/100 \quad (5.9)$$

$$X_h = X_l + PB \quad (5.10)$$

where

$C$ = present controller output (percent)
$C_o$ = initial controller output (percent)
$E$ = present controller error (percent)
$E_o$ = initial controller error (percent)
$t$ = present time for integral (seconds)
$t_o$ = initial time for integral (seconds)
$T_i$ = integral time for reset (seconds/repeat)
$PB$ = proportional band (percent)
$X_h$ = distance to *PB* high limit (percent)*
$X_l$ = distance to *PB* low limit (percent)*
$C_h$ = control valve stroke high limit (percent)
$C_l$ = control valve stroke low limit (percent)

Controllers with the antireset windup option limit the reset contribution so that $C$ cannot be greater than $C_h$ or less than $C_l$. The values of $C_h$ and $C_l$ are typically set by potentiometer adjustments. $C_h$ should be less than the maximum signal and $C_l$ should be greater than the minimum signal to which the control valve can respond. When $C$ is equal to $C_h$, $X_h$ is zero and $X_l$ is equal to the *PB*. When $C$ is equal to $C_l$, $X_l$ is zero and $X_h$ is equal to the *PB*. Thus the proportional band is located adjacent to the measurement on the side toward the set point so that proportional action starts as soon as the error reverses direction (not sign). Controllers without the antireset windup option can have a $C$ greater than $C_h$ or a $C$ less than $C_l$. When $C$ is greater than $C_h$, $X_h$ is negative whereas $X_l$ is equal to $X_h - PB$. Proportional action will not occur until after the error reverses direction and changes by an amount equal to $X_h$. When $C$ is less than $C_l$, $X_l$ is positive while $X_h$ is equal to $X_l + PB$. Porportional action will not occur until after the error reverses direction and changes by an amount equal to $X_l$. If the dis-

* Distance from measurement per Figure 5.1.

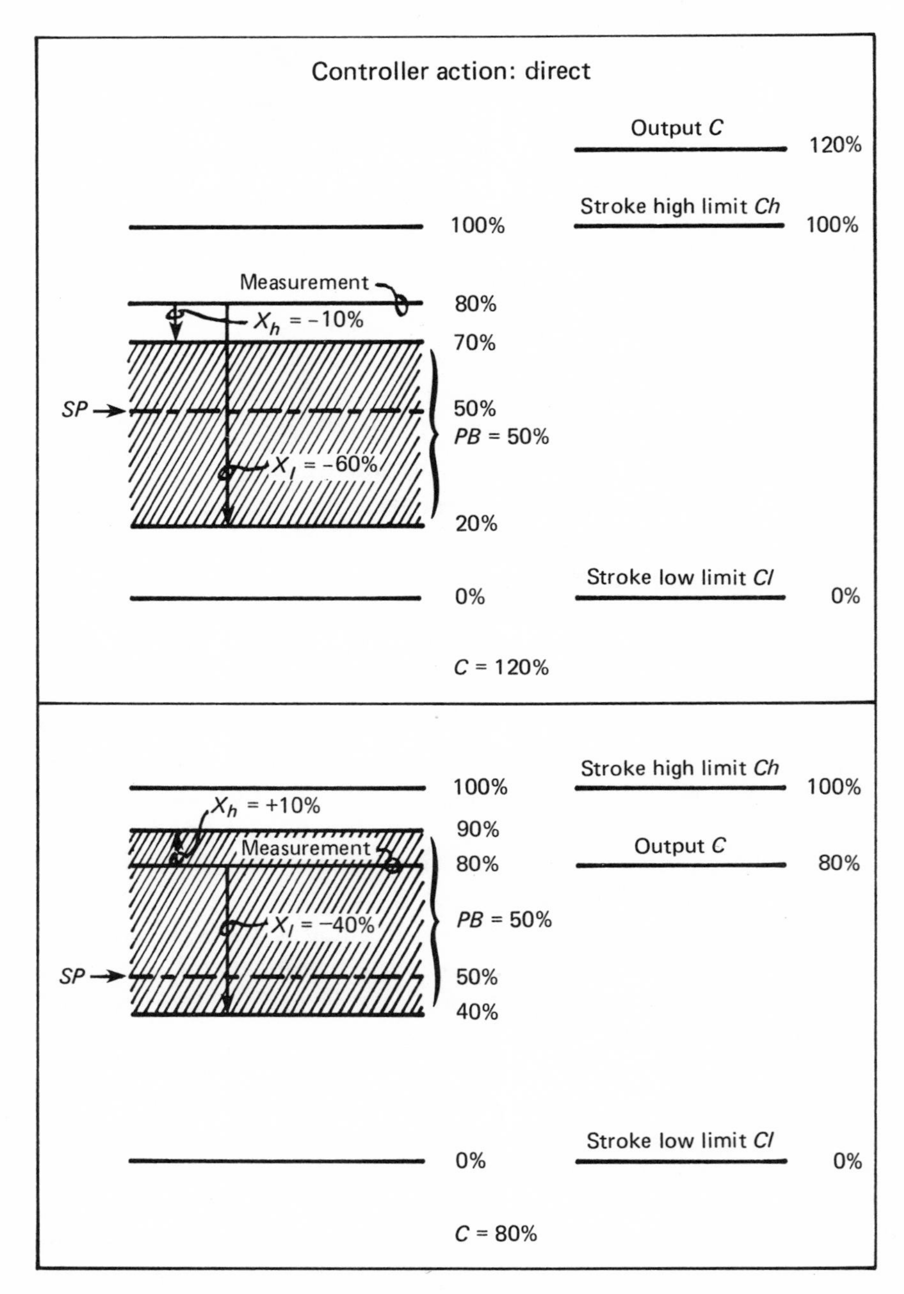


FIGURE 5.4
Location of proportional band for controller without ~~multi~~ anti reset windup.

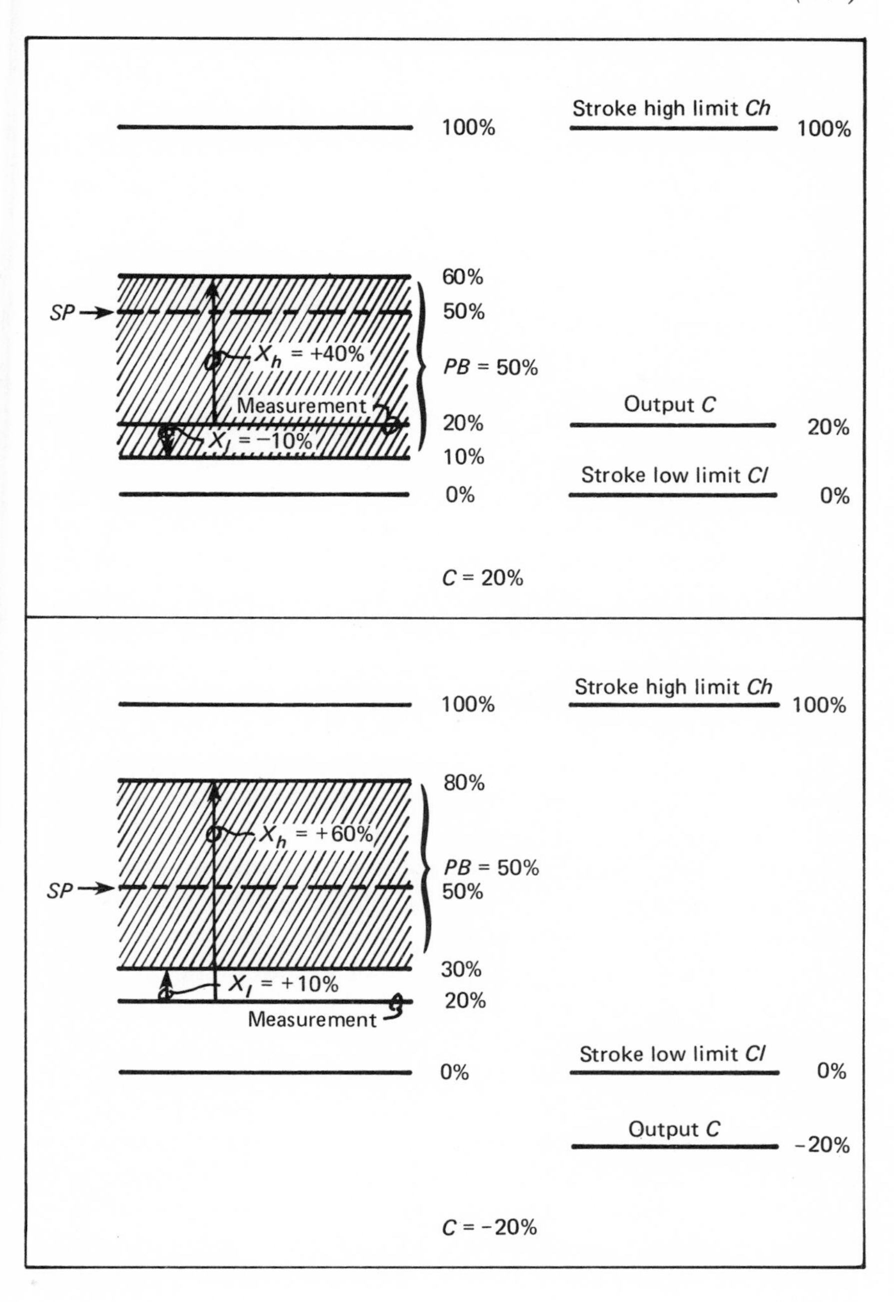
Stroke high limit Ch
100%
100%
60%
SP
50%
Xh = +40%
PB = 50%
Measurement
Output C
20%
20%
Xl = −10%
10%
Stroke low limit Cl
0%
0%
C = 20%
Stroke high limit Ch
100%
100%
80%
Xh = +60%
PB = 50%
SP
50%
30%
Xl = +10%
20%
Measurement
Stroke low limit Cl
0%
0%
Output C
−20%
C = −20%

placement of the proportional band indicated by $X_h$ for $C > C_h$ and by $X_l$ for $C < C_l$ is greater than the distance to the set point, the control valve will not move off its upper or lower limit until the error changes sign (the measurement crosses the set point and the reset contribution to the output changes sign). Figure 5.4 shows how the location of the proportional band shifts for various values of $C$ for controllers without the antireset windup option.

Some digital controllers have a set of adjustable output and antireset windup limits. The output limits are normally set to match the control valve stroke limits but larger than the antireset windup output limits. Between the output and antireset windup limits, the reset action is many times larger than selected so that the controller action rapidly unwinds. Proportional action does not take place until the measurement is within the proportional band centered about the set point. This type of proportional action is particularly useful for pressure control of vent valves where one does not want to pop the vent valve open for a large pressure increase, unless the increase approaches the set point. This type of control action is undesirable for compressor surge control because it delays opening of the surge control valve until the surge measurement is close to the surge line (the prestroke dead time of a closed surge control valve is significant). Proportional-only controllers also provide this action since their proportional band is centered around the set point. Some digital proportional plus manual bias controllers will shift the proportional band to the side closest to the measurement after switching from manual to automatic.

Supervisory controllers in general-purpose computers usually use a velocity instead of a positional algorithm to protect against reset windup and bumping the output during manual to automatic transfer and computer failure. The velocity algorithm computes a change in controller output whereas the positional algorithm computes a total or summed output. The change in supervisory controller output, which is the change in set point of the secondary controller, is usually represented by pulses with a logic signal for direction. If the error reverses direction and the proportional contribution is greater than the reset contribution to the output for that computation interval, the logic signal reverses the direction of the pulses. The set point would respond immediately to the change in direction of pulses, if limits were placed on the set point. (If limits were not placed on the set point, so that it was driven beyond the usable range of the secondary controller, then a form of reset windup could still occur.) When the computer is started or the supervisory loop is switched from manual to automatic, the pulses provide the incremental change from the last set point of the secondary controller. If the general-purpose computer fails, the pulses stop and the secondary controller set point is frozen at its last value. The velocity algorithm should be modified or replaced with a positional algorithm for direct digital control (DDC) of slow control valves because the velocity limiting of the valve can result in offset or wandering of the valve stroke because of saturation from noise (Bristol, 1977).

The override controller uses external feedback to prevent reset windup when the controller output is not chosen by a signal selector. In the controller by one manufacturer, the output is saturated when the internal feedback differs from the external feedback (when the controller output is not selected). If the measurement crosses set point, the controller output rapidly moves to saturation at the opposite end of the scale. When the internal feedback becomes equal to the external feedback, the controller is selected and normal PID action resumes. Thus the controller output indication will jump to the scale extreme when not selected, and to the output value last sent to the valve when selected. Even though the controller output indication jumps up and down the scale, the output to the valve is smooth. Some controllers ensure that selection of a controller does not occur until the measurement has crossed set point and that reset windup does not occur at either output extreme. (The reset contribution is set to zero when the controller output is at its limit.)

# CHAPTER VI
# *Effect of Measurement Dynamics*

## *6.1 Measurement Time Constant*

All time constants that exist between the controlled process variable and the controller input signal can be classified as measurement time constants. Sensors, transmitters, transducers, and filters usually have a time response that can be described by one or more time constants. If the measurement time constant is not the largest in the loop, it should be converted into equivalent dead time via Equations 4.1 and 4.2 ($TC_m = TC1$) and added to the total loop dead time for use in the equations in Chapter 4. If the measurement time constant is the largest time constant in the loop, all the other time constants in the loop should be converted to equivalent dead time via Equations 4.1 and 4.2 ($TC_m = TC2$) and added to any pure dead times for use in the equations in Chapter 4. If the measurement time constant is the largest time constant in the loop, it should be used as the loop time constant (*TC*) in the equations in Chapter 4 and the peak and accumulated errors should be multiplied by the ratio of the measurement to process time constant. These then are the errors in the controlled variable. The errors in the measured variable are equal to the calculated values without the ratio as a correction factor. The measured variable errors are much smaller than the controlled variable errors due to the filtering effect of the large measurement time constant.

Simulation results show that the loop period can be approximated by Equation 6.1 for self-regulating processes, if the measurement time constant is greater than 5 percent of the loop dead time.

For a self-regulating process:

$$T_c = 4*\left[1 + \left[\frac{TC_m}{TD}\right]^{0.5}\right]*TD \tag{6.1}$$

where

$T_c$ = loop period
$TC_m$ = largest measurement time constant
$TD$ = control loop dead time

$T_c$ is not the ultimate period, but it is the loop period with the existing reset and rate time settings. Also, the oscillations may be growing or decaying in magnitude, depending on the ratio of $TC_m$ to $TD$.

The addition of a measurement time constant greater than the loop dead time will cause the loop to go unstable unless it is retuned. New controller mode settings should be estimated by substitution of the new ultimate period into the equations in Chapter 4. If the time constant $TC_m$ is greater than 100 times the loop dead time $TD$, the measured oscillations for full-scale process oscillations are so small the operator may not be able to recognize loop instability (see Figure 6.1). For example, if the loop dead time is 0.2 minute and the measurement time constant is 20.0 minutes, the measured oscillations' amplitude will be 3.5 percent when the process oscillations' amplitude is 50.0 percent. The operator might mistake these measurement oscillations for noise.

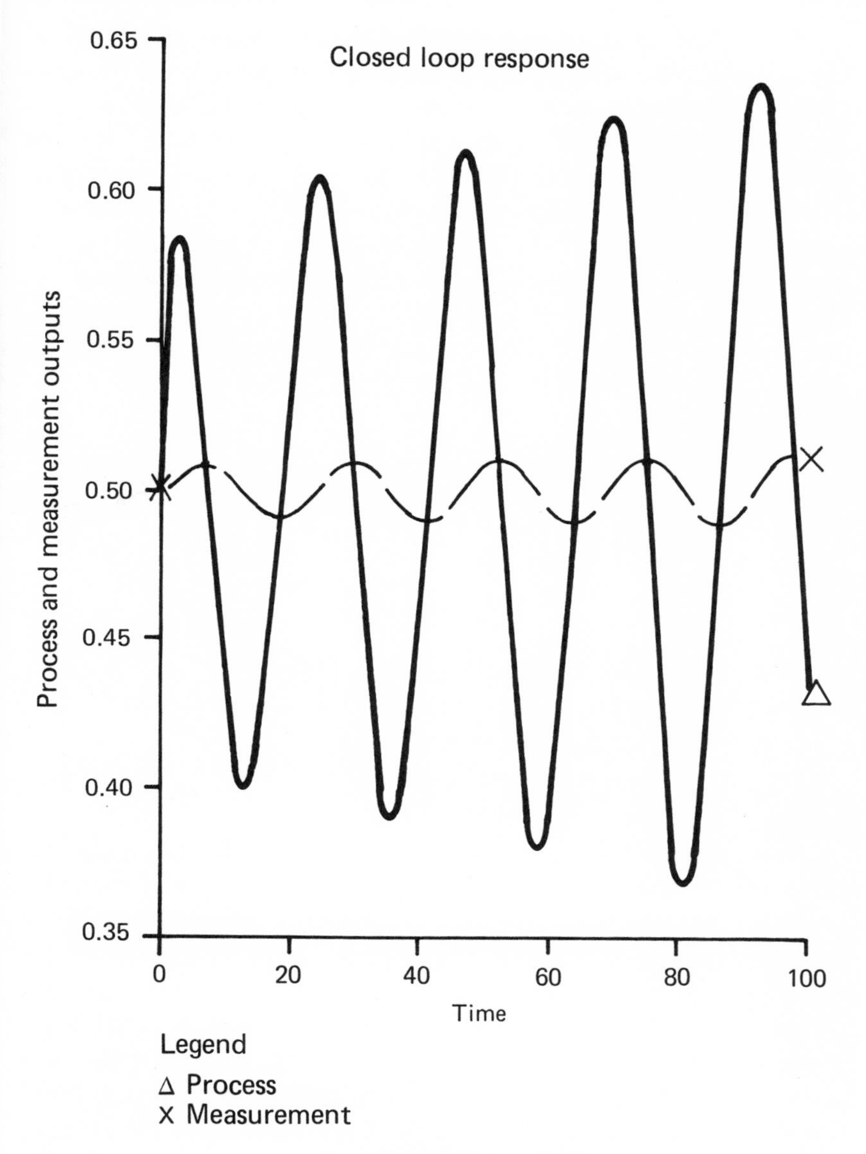


FIGURE 6.1
Closed loop response for large measurement time constant ($TC_m$ = 100*$TD$ for self-regulating process).

The signal used for local display, alarm, interlock, and communication over the data highway to the console or general-purpose computer in a distributed system is the filtered measurement signal. If a large time constant is selected for the digital filter (typically during the night shift), the alarms, interlocks, and displays may not be able to detect loop instability until a safety hazard has developed. The attenuation of the amplitude of the process oscillations may be estimated by Equation 6.2 for self-regulating processes when the measurement is linear and its time constant is greater than twice the dead time.

For $TC_m > 2*TD$:

$$A_m = \frac{A_p * T_c}{TC_m * 2\pi} \tag{6.2}$$

where

$A_m$ = amplitude of the measured oscillations
$A_p$ = amplitude of the process oscillations
$T_c$ = loop period from Equation 6.1
$TC_m$ = measurement time constant

Since the set point is usually not centered in the measurement range, the filtered measurement will drift away from set point to the average value of the unfiltered oscillations. If the measurement is nonlinear, the filtered measurement will not match the filtered process even though the measurement and process time constants are equal. For example, if a pH measurement filter is added to an in-line system and the time constant is set equal to the time constant of the downstream tank, the filtered in-line pH measurement amplitude will be smaller than the tank pH measurement amplitude and the longer loop period will decrease the effectiveness of the filtering action of the tank's volume (Equation 6.2 can be used to determine the effectiveness of the tank's filtering action if concentration units are used for both amplitudes and the tank's time constant is used for $TC_m$). On some loops, the noise is so severe that the degradation in loop performance due to noise is greater than the degradation in loop performance due to the increase in loop period by the filter. If the period of the noise is less than 5 percent of the loop dead time, then a filter time constant can be carefully selected to give a negligible increase in loop period (see Chapter 8).

Simulation results show that the loop period can be approximated by Equation 6.3 for an integrating process if the measurement time constant is greater than 5 percent of the dead time.

For an integrating process:

$$T_c = 4*\left[1 + \left[\frac{TC_p}{TD}\right]^{0.65} + \frac{TC_m}{TD}\right]*TD \qquad (6.3)$$

where

$T_c$ = loop period
$TC_p$ = process time constant
$TD$ = loop dead time
$TC_m$ = measurement time constant

Equation 6.3 shows that the new loop period is equal to the original ultimate period for an integrating process plus four times the measurement time constant. All of the measurement time constant is converted to dead time. For a given increase in measurement time constant, the increase in loop period is greater for the integrating process than for the self-regulating process. For a given increase in loop period, the increase in peak and accumulated errors is greater for the integrating process than for the self-regulating process (the peak error is proportional to the ultimate period squared and the accumulated error is proportional to the ultimate period cubed for an integrating process). The attenuation in process oscillations due to the measurement time constant is much less for an integrating process than for a self-regulating process, since the numerator increases as fast as the denominator in Equation 6.2 for an integrating process. If a loop with an integrating process goes unstable due to a large measurement time constant, the displays, alarms, and interlocks will see the instability. However there will be a time delay between the measured and process oscillations. This time delay will be approximately equal to the measurement time constant, $TC_m$, if the measurement time constant is not greater than ten times the dead time.

A simple expression for the loop period of a runaway process for a large measurement time constant has not been developed to date by the author. However Luyben has developed an equation to predict the effect of a measurement time constant and a heat removal time constant on the ratio of maximum to minimum proportional band if the effect of dead time is neglected (Luyben & Melcic, 1978). It is not good practice to ignore the effect of dead time, as is evident in previous equations for the ultimate period. However Equation 6.4 can be used to predict the maximum ratio of maximum to minimum proportional bands. Dead time will make the actual ratio smaller. Luyben states that a minimum ratio of about 15 is necessary to avoid instability for changing process gains and dynamics in chemical plants.

$$R_c = 1 + (1 - R_m - R_h)^* \left[ \frac{R_m + R_h}{R_m{}^* R_h} - 1 \right] \tag{6.4}$$

where

$R_c$ = ratio of maximum to minimum proportional band

$R_m$ = ratio of measurement to runaway time constant $\dfrac{TC_m}{TC'}$

$R_h$ = ratio of heat removal to runaway time constant $\dfrac{TC_h}{TC'}$

The window of allowable proportional bands decreases as either the measurement or heat removal time constant increases for a given runaway time constant $TC'$ (see Figure 6.2). The window closes if the summation of the measurement and heat removal time constant ratios is greater than 1 ($R_m + R_h > 1.0$) or the product of these time constants is larger than the sum ($R_m{}^*R_h > R_m + R_h$). Once the window closes, the loop is unstable for all controller mode settings and all known feedback control algorithms.

Figure 6.3 shows the effect of various transmitter time constants on the precipitous drop in flow and the flow oscillations during surge. If the time constant is very large, such as that found in a pneumatic transmitter with high damping ($TC_m$ = 16 seconds), the operator and control system will not even realize a surge has occurred owing to the extensive filtering by the transmitter.

The measurement time constant for a pH electrode depends upon the direction of the pH change, the magnitude of the pH change, the degree of buffering, fluid velocity, and the type of electrode flow cell. Table 6.1 is based on data from Giusti and Hougen (1961) and Green and Field (1964).

Table 6.1 shows that the electrode time constant for a negative pH change is about twice that for a positive pH change. Low fluid velocities can cause large electrode time constants for negative pH changes. Fluid velocities in flow cells or past submersion assemblies in vessels are usually less than 1 fps.

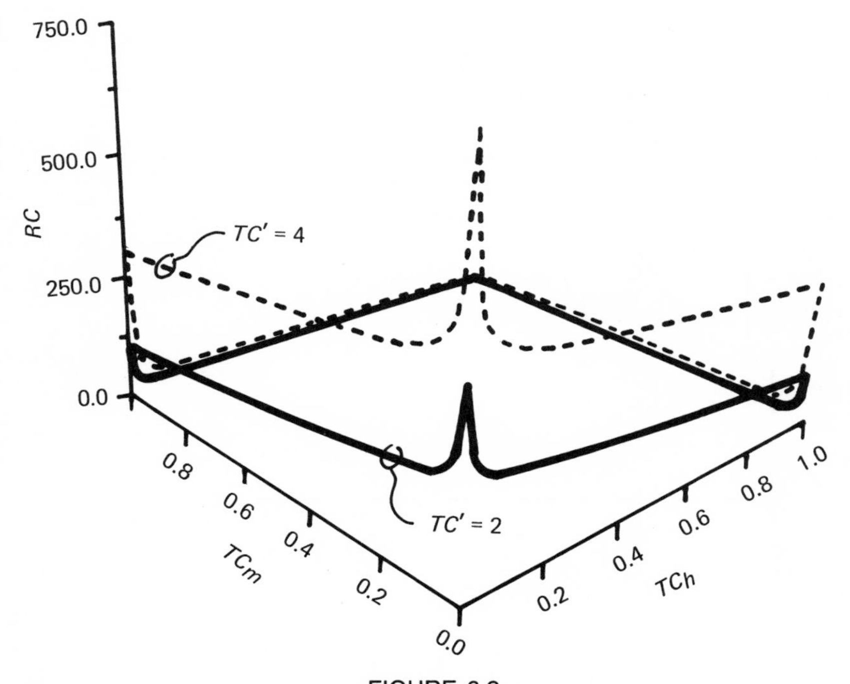


FIGURE 6.2
Ratio of maximum to minimum proportional band for a runaway process.

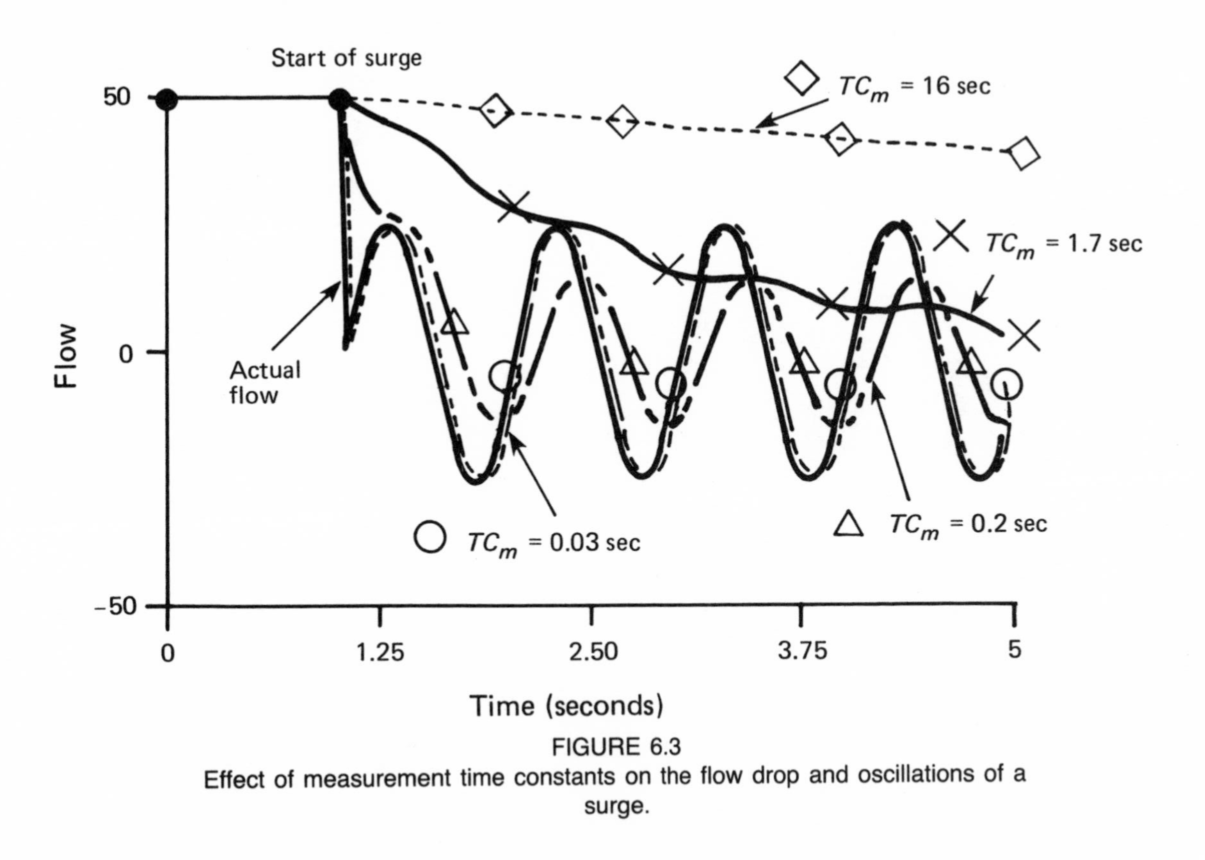


FIGURE 6.3
Effect of measurement time constants on the flow drop and oscillations of a surge.

If the flow cell volume is not negligible, another measurement time constant exists that is equal to the volume of the cell divided by the sample flow (complete mixing of the cell volume due to flow turbulence can be assumed). If the electrode has a coating, the measurement time constant will naturally increase. Polymer, tar, and oil coatings can be decreased by high fluid velocities (7 to 10 fps) and crystal and particle coatings can be reduced by ultrasonic cleaners. An injector type of electrode assembly that allows electrode replacement in a pressurized connection and insertion into high-velocity streams in agitated vessels or pipelines has been proven at many installations. Flow cells or pots should be avoided unless the additional volume is to be used instead of a filter to reduce excessive measurement noise.

Temperature sensors in thermowells have two measurement time constants whose sizes depend upon the fluid type, fluid velocity, fluid temperature, annular clearance, and annular fill. Table 6.2 presents calculated time constants for thermocouples in thermowells (Buckley, 1979). These data can be used for other types of sensors in thermowells because the effect of sensor mass and diameter is negligible compared with the effect of fluid conditions and annulus characteristics. If the thermowell has a glass coating, the larger time constant should be increased

TABLE 6.1
pH Electrode Time Constants for Negligible Flow Cell Volume

| *Direction* | *Magnitude, pH* | *Buffering* | *Velocity, fps* | *Time Constant, seconds* |
|---|---|---|---|---|
| Positive | 0.5 | No | 5 | 1.2 |
| Negative | 0.5 | No | 5 | 2.8 |
| Positive | 1.0 | Yes | 5 | 0.25 |
| Negative | 1.0 | Yes | 5 | 0.5 |
| Positive | 1.5 | No | 5 | 1.8 |
| Negative | 1.5 | No | 5 | 6.2 |
| Positive | 3.0 | Yes | 4 | 0.75 |
| Negative | 3.0 | Yes | 4 | 1.5 |
| Positive | 3.0 | Yes | 2 | 1.5 |
| Negative | 3.0 | Yes | 2 | 3 |
| Positive | 3.0 | Yes | 1 | 2 |
| Negative | 3.0 | Yes | 1 | 4 |
| Positive | 3.0 | Yes | 0.5 | 3 |
| Negative | 3.0 | Yes | 0.5 | 12 |

by 30 percent or more. Bare temperature sensors have single time constants that depend on sensor construction and fluid conditions. Table 6.3 presents time constants from the literature for bare temperature sensors in liquids at velocities from 1 to 3 fps (Kardos, 1977). Response tests at Oak Ridge National Laboratory showed that the tightness or looseness of the MgO insulation in a sheathed thermocouple had a much greater effect on the bare element time constant than did junction mass and closure weld mass. The time constant was decreased by up to 56 percent by the swaging of the sheath tip to pack the MgO insulation tighter (Carroll et al., 1982).

A technique has been developed to estimate the effect of fluid conditions a priori and accurately measure a thermocouple or resistance temperature detector (RTD) response in situ. Tests show that it is difficult accurately to estimate the time constant of an installed temperature sensor. A better approach is to make a measurement of the response of the type of sensor used at one set of fluid conditions and then estimate the effect of the fluid conditions for various applications. An increase in fluid velocity increases the surface heat transfer coefficient. Thus the

TABLE 6.2
Thermowell and Thermocouple Assembly Time Constants

| *Fluid Type** | *Fluid Velocity, fps* | *Annular Clearance, inch* | *Annular Fill* | *Time Constants, seconds* |
|---|---|---|---|---|
| Gas | 5 | 0.04 | Air | 107 and 49 |
| Gas | 50 | 0.04 | Air | 93 and 14 |
| Gas | 152 | 0.04 | Air | 92 and 8 |
| Gas | 300 | 0.04 | Air | 92 and 5 |
| Gas | 152 | 0.04 | Oil | 22 and 7 |
| Gas | 152 | 0.04 | Mercury | 17 and 8 |
| Gas | 152 | 0.02 | Air | 52 and 9 |
| Gas | 152 | 0.005 | Air | 17 and 8 |
| Liquid | 0.01 | 0.01 | Air | 62 and 17 |
| Liquid | 0.1 | 0.01 | Air | 32 and 10 |
| Liquid | 1 | 0.01 | Air | 26 and 4 |
| Liquid | 10 | 0.01 | Air | 25 and 2 |
| Liquid | 10 | 0.01 | Oil | 7 and 2 |
| Liquid | 10 | 0.01 | Mercury | 2 and 0.2 |
| Liquid | 10 | 0.055 | Air | 228 and 1 |
| Liquid | 10 | 0.005 | Air | 4 and 1 |

* The gas is saturated steam and the liquid is organic.

large time constant typically changes with the fluid velocity to the −0.6 power. The effect of fluid temperature is more difficult to estimate since it affects the conductivity of the sensor as well as the surface heat transfer coefficient. The time constant increases with temperature for most sensors. If the fluid conditions are unpredictable, sensor fouling occurs, or accurate knowledge of the sensor dynamics is necessary for an advanced control strategy, then the sensor time constant should be calculated from a test with the sensor in service (in situ). The loop current step response (LCSR) method can accurately calculate the time constant of installed thermocouple or resistance temperature detectors by applying a small alternating or direct current to the sensor leads to heat the sensing element internally and analyzing the resulting temperature transient (Kerlin et al., 1982).

The measurement time constants of electronic temperature transmitters and magnetic flow transmitters are negligible. The measurement time constants for other transmitters depend on the type, manufacturer, model, and signal. The more commonly used transmitters are listed in Table 6.4. If a particular transmitter of interest does not appear there, the manufacturer's local representative should be able to get the measurement time constant from the factory. The *ISA Transducer Compendium*

TABLE 6.3
Bare Temperature Element Time Constants

| *Bare Element Type* | *Time Constant, seconds* |
|---|---|
| Thermocouples: | |
| 1/8 inch sheathed and grounded | 0.3 |
| 1/4 inch sheathed and insulated | 4.5 |
| 1/4 inch sheathed and grounded | 1.7 |
| 1/4 inch sheathed and exposed loop | 0.1 |
| Resistance temperature detectors (RTD): | |
| 1/16 inch | 0.8 |
| 1/8 inch | 1.2 |
| 1/4 inch | 5.5 |
| 1/4 inch dual element | 8.0 |
| Mercury-filled bulb: | |
| 1/4 inch | 1.6 |
| 3/8 inch | 2.5 |
| 3/4 inch | 6.5 |

is a good source of data on measurement time constants and errors.

The time constant and dead time for $\frac{1}{4}$- and $\frac{3}{8}$-inch pneumatic tubing into a bellows termination for a 3–15 psi step change can be estimated from Table 6.5 (Bradner, 1949). The dead time was estimated to be approximately 20 percent of the total lag time to reach 63 percent of the total change. The table is useful for estimating the effect of tubing length from an I/P transducer to a positioner or booster. Sometimes the I/P is remotely mounted from the actuator for maintenance accessibility or to avoid excessive vibration, temperature, or corrosion.

## 6.2 Measurement Dead Time

Measurement dead time can originate from multiple time constants in series, transportation delay, and sampling. Measurement dead time resulting from multiple time constants can be estimated by Equations 4.1 and 4.2 where $TC2$ is the largest time constant in the loop ($TC2$ is usually the process time constant $TC_p$). If $TC2$ is greater than 20 times $TC1$, the measurement dead time $TD_m$ is approximately equal to $TC_m$. The transportation delay can be estimated by Equation E1 from Appendix E where the ratio $F/F_a$ is 1.5 for laminar flow and 3.0 for turbulent flow. The remaining time constant per Equation E2 may also be converted to dead time per Equations 4.1 and 4.2. If the remaining time constant is converted to dead time, a simpler approximation that the measurement dead time is equal to the residence time (volume divided by throughput flow) can be used. Sampling with a zero-order hold creates a dead time equal to one half of the sample time. If the measurement output is updated at the end of the sample period, then an additional dead time equal to the sample time is created. Thus the measurement dead time for instruments such as the chromatograph is equal to 1.5 times the sample time.

## TABLE 6.4
## Miscellaneous Electronic Transmitter Time Constants

| Transmitter Type | Manufacturer and Model | Time Constant, seconds |
|---|---|---|
| Differential pressure | Rosemount 1151DP | 0.2–1.7* |
| Gage pressure | Rosemount 1151GP | 0.2–1.7* |
| Absolute pressure | Rosemount 1151AP | 0.2–1.7* |
| Flange-mounted level | Rosemount 1151LL | 0.2–1.7* |
| Differential pressure | Foxboro 823DP | 0.2–1.6† |
| Gage pressure | Foxboro 823GM | 0.2–1.6† |
| Absolute pressure | Foxboro 823AM | 0.2–1.6† |
| Flange-mounted level | Foxboro E17 | 0.3 |
| Diaphragm seal d/p | Foxboro E13DMP | 1.6 (M capsul) |
| Diaphragm seal d/p | Foxboro E13DMP | 0.5 (H capsul) |
| Turbine flowmeter | Foxboro 81A | 0.03 maximum |
| Transmitting rotameter | Wallace & Tiernan | 0.2 |
| Speed (magnetic pickup) | Dynalco SS | 0.04 (~2,000 Hz) |
| Speed (magnetic pickup) | Dynalco SS | 0.2 (~400 Hz) |
| Speed (magnetic pickup) | Dynalco SS | 0.8 (~80 Hz) |
| Speed (magnetic pickup) | Dynalco SS | 3.5 (~15 Hz) |
| Vortex flowmeter | Fischer & Porter 10LV2 | 2.5 |
| Nuclear density gage | Texas Nuclear SGH | 15–300* |
| Nuclear density gage | Kay-Ray | 2.2–26* |
| Nuclear level gage | Kay-Ray | 0.4–13* |

* Time constant is adjustable.
† Time constant is selected by three-position jumper.

## TABLE 6.5
## Pneumatic Tubing Dead Times and Time Constants

| Tubing ID, inch | Tubing Length, feet | Dead Time, seconds | Time Constant, seconds |
|---|---|---|---|
| 0.188 | 50 | 0.06 | 0.24 |
| 0.188 | 100 | 0.12 | 0.48 |
| 0.188 | 200 | 0.36 | 1.44 |
| 0.188 | 300 | 0.58 | 2.32 |
| 0.188 | 400 | 0.84 | 4.20 |
| 0.188 | 500 | 1.20 | 4.80 |
| 0.188 | 1000 | 3.80 | 15.20 |
| ~~0.305~~ 0.188 | 2000 | 12.0 | 48.00 |
| 0.305 | 50 | 0.04 | 0.14 |
| 0.305 | 100 | 0.06 | 0.26 |
| 0.305 | 200 | 0.16 | 0.64 |
| 0.305 | 300 | 0.28 | 1.12 |
| 0.305 | 400 | 0.44 | 1.76 |
| 0.305 | 500 | 0.64 | 2.56 |
| 0.305 | 1000 | 2.00 | 8.00 |
| 0.305 | 2000 | 6.00 | 24.0 |

For chromatographs:

$$TD_m = 1.5 * T_s \tag{6.5}$$

where

$TD_m$ = measurement dead time
$T_s$ = chromatograph sample time

Since chromatographs can have sample times of 3 to 30 minutes, the loop performance is seriously degraded for self-regulating loops unless the process time constant $TC$ is exceptionally large. Integrating loops will be unstable unless the integrator gain is exceptionally small and runaway loops will be unstable unless the runaway time constant, $TC'$, is exceptionally large. If the chromatograph provides the only process measurement, then it is difficult to conduct an open loop test graphically to determine the process time constant and dead time. The resolution of the graphical method is limited to the chromatograph sample time unless multiple response tests are made and the offset between the disturbance and sample start times is varied. The mass spectrometer can provide an analysis in 10 to 40 seconds and should be considered for control when the performance of the loop is important (Weiss, 1977).

## 6.3 Measurement Accuracy and Rangeability

Measurement accuracy is the limits within which the measured value of a process variable might vary relative to its true value. It includes measurement nonlinearity, hysteresis, and repeatability errors. Measurement nonlinearity is the maximum deviation from a straight line drawn from 0 to 100 percent of scale. Measurement hysteresis is the maximum difference of measurements when approached from 0 percent and from 100 percent of scale. Measurement repeatability is the maximum deviation of measurements for identical values of the process variable approached from the same direction occurring at different times (Liptak, 1969).

Measurement nonlinearity, which is usually the greatest contributor to measurement error, can be reduced either by operation close to one set point that is biased for the nonlinearity error or by signal linearization. For example, the vortex flowmeter has a 1 percent nonlinearity error and a 0.1 percent repeatability error in the meter coefficient. Most of the nonlinearity error can be eliminated by use of a polynomial of the measured frequency in hertz divided by the fluid kinematic viscosity (McMillan, 1981a). The resolution error of analog measurements is negligible (resolution is the smallest change in the process variable that produces a detectable change in the measurement signal). The resolution error of digital measurements, which is set by the number of bits in the transmitter or the sample time and the rate of change of the process variable, may be the limiting factor in digital transmitter accuracy.

For digital transmitters whose resolution is limited by number of bits:

$$E_r = P/(2^{Nb}) \qquad (6.6)$$

For digital transmitters whose resolution is limited by sample time:

$$E_r = \frac{\Delta P}{\Delta T} * T_s \qquad (6.7)$$

where

$E_r$ = resolution error in process units
$P$ = full-scale value of the process variable being measured
$N_b$ = number of bits in the digital transmitter
$T_s$ = sample time of the digital transmitter
$\frac{\Delta P}{\Delta T}$ = rate of change of the process variable being measured

The standard deviation of the controlled process variable, $E_p$, can be estimated as the square root of the sum of control error standard deviation, $E_c$, squared, and the individual measurement errors, $E_m$, squared (Stout, 1976).

$$E_p = (E_c^2 + E_{m1}^2 + E_{m2}^2)^{0.5} \tag{6.8}$$

where

$E_p$ = standard deviation of the controlled process variable
$E_c$ = standard deviation of control errors
$E_{m1}$ = measurement error for instrument no. 1
$E_{m2}$ = measurement error for instrument no. 2

The standard deviation of the control errors, $E_c$, can be estimated for a given peak error, ultimate period, and disturbance time interval. The deviation, $E_c$, is approximately one fourth of the peak error for an ultimate period that is approximately one third of the disturbance time interval. For this condition, the reduction of the peak error below the measurement error has an insignificant effect on the standard deviation of the process variable. In the selection of a measurement location (e.g., distillation column tray number for temperature control), it may be more important to maximize the measurement gain than minimize the loop dead time because $E_m$ is larger than $E_c$. (The tradeoff should be evaluated.) A larger measurement gain translates to a smaller standard deviation of the actual controlled process variable (e.g., distillation column composition for tray temperature control).

Measurement rangeability is the ratio of the maximum to the minimum measurement signal that is still within the stated accuracy. The minimum signal may be limited by noise, nonlinearity, or resolution. Since flow measurements are frequently noisy and nonlinear, the rangeability of the flow transmitter should be checked during design, especially in pH loops. Since control valves typically have much larger rangeabilities than flow transmitters, the rangeability of the manipulated variable may be severely reduced in a cascade loop that has an inner flow loop. Magnetic flowmeters have the highest rangeability

(50:1) and differential head flowmeters have the lowest rangeability (4:1). The vortex flowmeter rangeability (normally 15:1) increases and approaches the magnetic flowmeter rangeability as the kinematic viscosity of the liquid decreases.

## 6.4 *Examples*

### 6.4.1 WASTE TREATMENT PH LOOP (SELF-REGULATING)

*Given:*

(Information from Example 4.6.1)

(a) Set point is 7 pH.
(b) Measurement range is 0 to 14 pH.
(c) Minimum influent flow is 10 gpm (*Fil*).
(d) Normal influent flow is 22 gpm (*Fin*).
(e) Maximum influent flow is 100 gpm (*Fih*).
(f) Influent concentration is 32 percent by weight HCl—10.17 normality ($C_i$).
(g) Influent disturbance is rapid 20 gpm increase in flow ($\Delta L$).
(h) Reagent concentration is 20 percent by weight NaOH—7.93 normality ($C_r$).
(i) Vertical tank liquid volume is 1000 gallons ($V$).
(j) Axial blade agitator diameter is 1 foot ($D$).
(k) Axial blade agitator speed is 120 rpm ($N_s$).
(l) Axial blade agitator discharge coefficient is 1 ($N_q$).

(New information)

(m) Volume of pipeline from tank to pH electrodes is 40 gallons ($V_e$).
(n) Recirculation turbulent flow in pipeline to electrodes is 15 gpm ($F_e$).

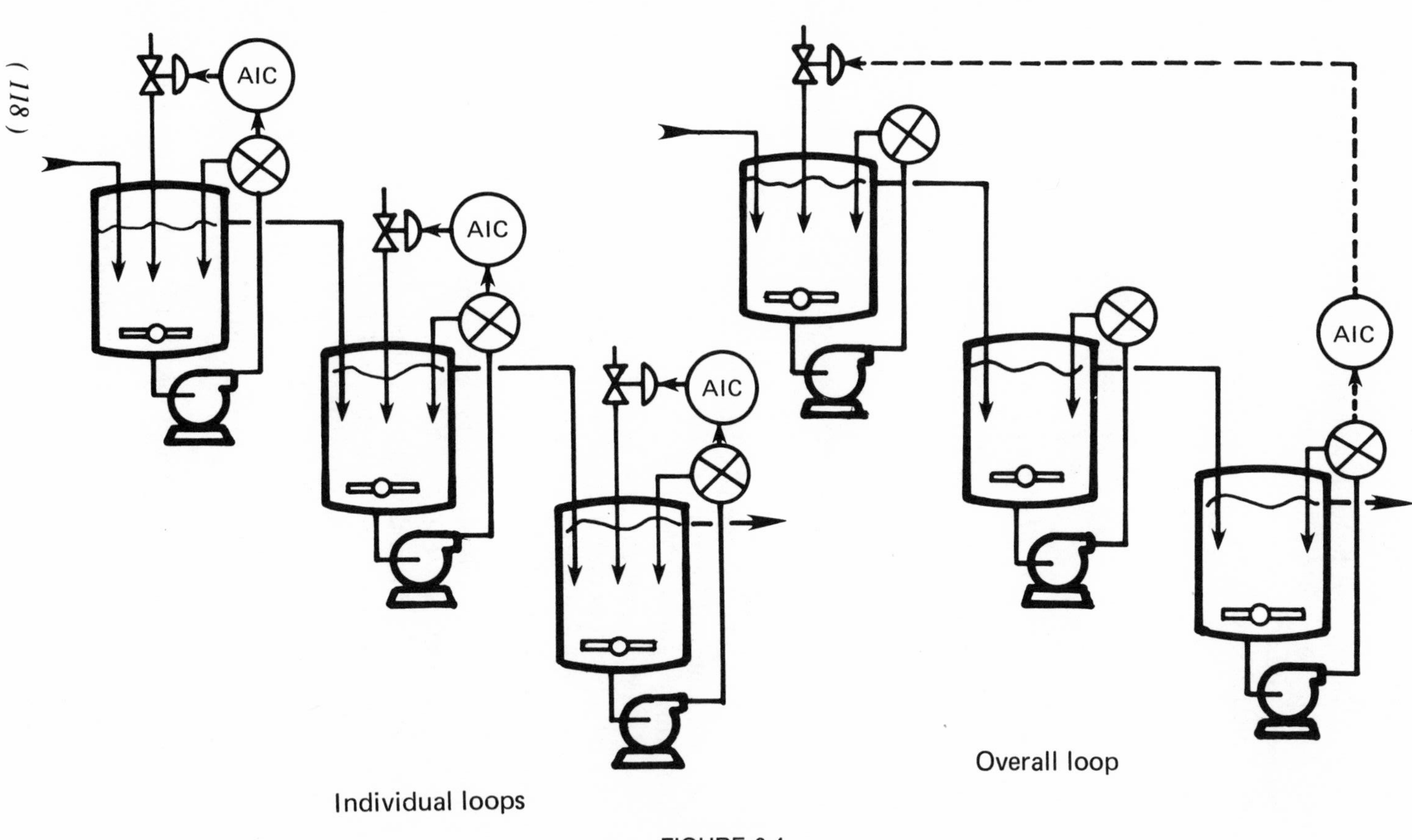


FIGURE 6.4
Waste treatment pH loop.

*Find:* the new peak and accumulated errors for the individual tank loops.

*Solution:*

(a) Calculate the new ultimate period for the individual loop:

$$TD_m = \left[\frac{V_e}{F_e + F_{ae}}\right] + Y^*TC_m$$

$$TD_m = \left[\frac{40}{15 + \frac{15}{3}}\right] + 1^*\frac{6.2}{60} \qquad \text{(E1 and 4.2)}$$

$$= 2.1 \text{ minutes}$$

(See Appendix E for the equation for the turbulent flow pipeline delay. See Table 6.1 for the electrode time constant for a large pH decrease.)

$$TD = TD_p + TD_m$$

$$TD = 1 + 2.1 = 3.1 \text{ minutes per tank}$$

$$TC = TC_p = 19 \text{ minutes per tank}$$

$$T_u = 2^*\left[1 + \left[\frac{TC}{TC + TD}\right]^{0.65}\right]^*TD \qquad (4.5)$$

$$T_u = 2^*\left[1 + \left[\frac{19}{19 + 3.1}\right]^{0.65}\right]^*3.1$$

$$= 12 \text{ minutes}$$

(b) Calculate the new peak and integrated errors in concentration units:

$$E_x = \left[\frac{1.1^*K_g^*T_u}{2\pi^*TC}\right]^*E_o \qquad (4.14)$$

$$E_x = \left[\frac{1.1^*1.5^*12}{2\pi^*19}\right]^*2$$

$$E_x = 0.17*2 = 0.3 \text{ normality for the first tank}$$

$$E_x = 0.17*0.3 = 0.05 \text{ normality for the second tank}$$

$$E_x = 0.17*0.05 = 0.009 \text{ normality for the third tank}$$

$$E_i = \frac{E_x}{1\ 1} *0.5*T_u \text{ (Equation 4.14 substituted into Equation 4.13)}$$

$$E_i = \frac{0.3}{1.1} *0.5*3.1$$

$$= 0.4 \text{ normality*minute for the first tank}$$

$$E_i = \frac{0.05}{1.1} *0.5*3.1$$

$$= 0.07 \text{ normality*minute for the second tank}$$

$$E_i = \frac{0.009}{1.1} *0.5*3.1$$

$$= 0.01 \text{ normality*minute for the third tank}$$

*Conclusions:* The pipeline transportation delay increases the dead time in the loop by a factor of 3. The peak error in concentration units increased by a factor of 3 for each tank for a total factor of 27 for the three tanks. The accumulated error in concentration units increased by a factor of 9 for each tank, for a total factor of 729 for the three tanks. The increase in pH errors will not be as great as a result of the exponential relationship between normality and pH. The electrode time constant was entirely converted to effective dead time, since the electrode time constant was less than 5 percent of the residence time. The additional error due to transportation delay is a common problem because the electrodes are frequently located in a recirculation pipeline instead of inside the tank for easier access for maintenance. Unfortunately the location chosen is near the reagent valves at a platform on the tank top that is the farthest point from the tank discharge nozzle (the pipeline volume, $V_e$, is large). The transportation delay can be reduced either by locating the electrodes closer to the pump or by increasing the recirculation flow.

## 6.4.2 BOILER FEEDWATER FLOW LOOP (SELF-REGULATING)

*Given*:

(Information from Example 4.6.2)

(a) Set point is 100,000 pph.
(b) Measurement range is 0 to 200,000 pph.
(c) Disturbance is rapid 20 percent increase in flow ($\Delta L$).
(d) Pipe diameter is 4 inches or 0.33 foot ($D_p$).
(e) Fluid velocity is 5 fps ($V_f$).
(f) Pipe friction factor is 0.01 ($C_p$).
(g) Pipe wall modulus of elasticity is 500,000,000 lb/ft$^2$ ($E_p$).
(h) Pipe wall thickness is 0.34 inch or 0.03 foot ($H$).
(i) Fluid density is 62 lb/ft$^3$ ($W$).
(j) Fluid modulus of elasticity is 5,000,000 lb/ft$^2$ ($E_f$).
(k) Acceleration due to gravity is 32 ft/(sec*sec) ($G$).
(l) Pipe length from valve to transmitter or discharge is 150 feet ($S_p$).
(m) Ratio of valve pressure drop to pipe pressure drop is 0.7 ($R$).

(New information)

(n) Flow transmitter is Rosemount 1151DP with time constant set at maximum.

*Find:* the new peak and accumulated errors.

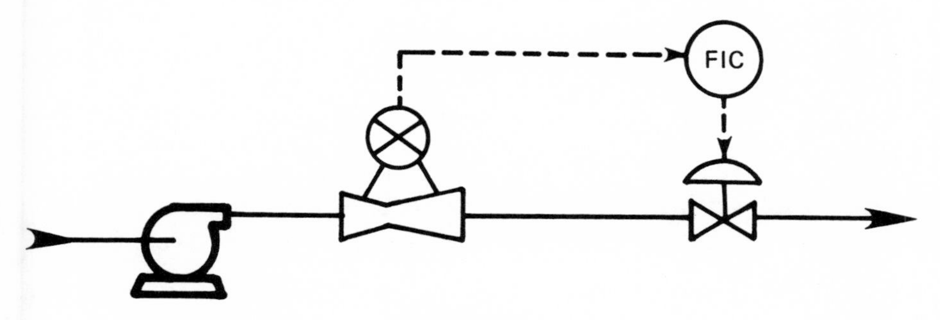


FIGURE 6.5
Boiler feedwater flow loop.

*Solution:*

(a) Calculate the new loop period:

$$T_u = 4*\left[1 + \left[\frac{TC}{TC + TD}\right]^{0.5}\right]*TD \qquad (4.5)$$

$$TC_m = TC2$$

$$TC_p = TC1$$

$$TD = Y*TC2 + 0.1 \qquad (4.2)$$

$$TD = 0.36*1 + 0.1 = 0.46 \text{ second}$$

$$T_u = 4*\left[1 + \left[\frac{1.7}{1.7 + 0.46}\right]^{0.5}\right]*0.46 \qquad (6.1)$$

$$T_u = 1.7 \text{ seconds}$$

(b) Calculate the new peak and accumulated errors:

$$E_x = \left[\frac{1.1*K_g*T_u}{2\pi*TC}\right]*E_o*\frac{TC_m}{TC_p} \qquad \text{modified (4.14)}$$

$$E_x = \left[\frac{1.1*1.5*1.7}{2\pi*1.7}\right]*10*\frac{1.7}{1}$$

$$E_x = 4.5 \text{ lb/sec} = 15910 \text{ pph}$$

$$E_i = \left[\frac{E_x}{1.1}\right]*T_i \qquad \text{(Equation 3.11 substituted into Equation 3.7)}$$

$$T_i = 0.8*T_u \qquad (4.7)$$

$$T_i = 0.8*1.7 = 1.4 \text{ seconds}$$

[Here 1.4 seconds is 43 repeats per minute, which is between the reset settings of 20 and 50 repeats per minute. Twenty repeats per minute should be used to avoid a reset cycle ($T_i$ = 3 seconds).]

$$E_i = \left[\frac{4.5}{1.1}\right]*3 = 12 \text{ pounds}$$

*Conclusions:* The loop period and peak error both increased by a factor of 4. The accumulated error would normally have increased by a factor of 16, but the reset setting resolution caused the accumulated error to increase by a factor of 24. The predicted errors are approaching those experienced in the field but are still somewhat smaller because the effects of valve dynamics and measurement noise have not been included.

### 6.4.3 BOILER DRUM LEVEL LOOP (INTEGRATING)

The addition of measurement dynamics has a negligible effect on the errors since the required proportional band is still far below the minimum proportional band available.

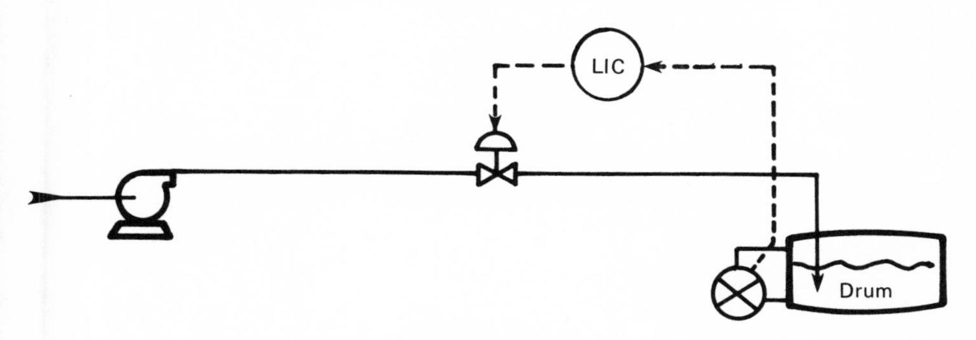


FIGURE 6.6
Boiler drum level loop.

### 6.4.4 FURNACE PRESSURE LOOP (PSEUDOINTEGRATING)

*Given:*

(Information from Example 4.6.4)

(a) Set point is 5 inches w.c. gage.
(b) Measurement range is 0 to 10 inches w.c. gage.
(c) Furnace volume is 10,000 ft$^3$ ($V_f$).
(d) Quench volume is 1000 ft$^3$ ($V_q$).
(e) Scrubber volume is 1000 ft$^3$ ($V_s$).
(f) Furnace inlet flow resistance pressure drop is 2.5 inches w.c. ($\Delta P_f$).

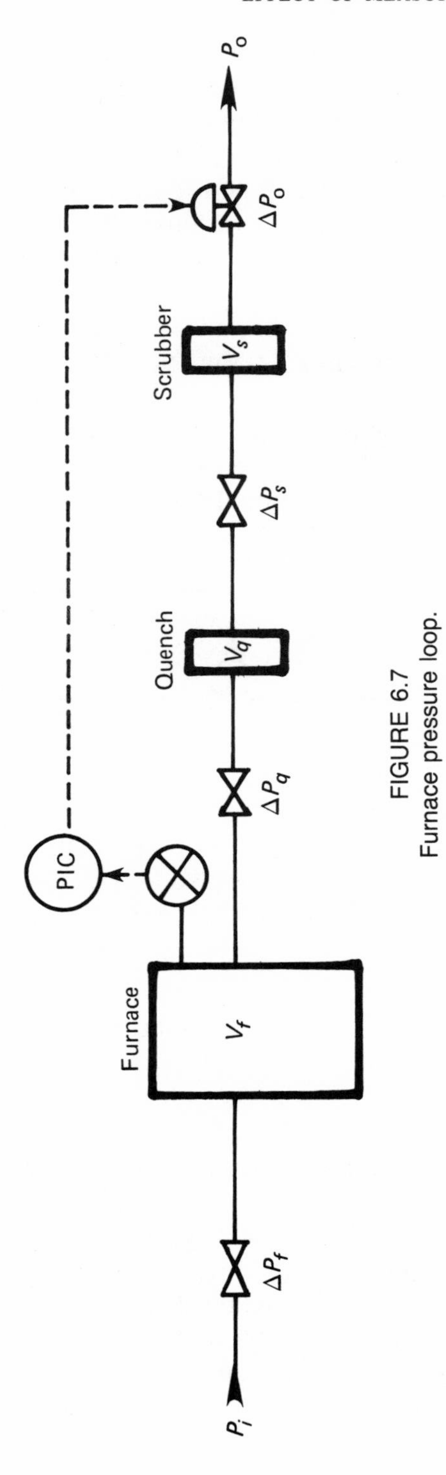


FIGURE 6.7
Furnace pressure loop.

(g) Quench inlet flow resistance pressure drop is 5 inches w.c. ($\Delta P_q$).
(h) Scrubber inlet flow resistance pressure drop is 10 inches w.c. ($\Delta P_s$).
(i) System outlet flow resistance pressure drop is 2.5 inches w.c. ($\Delta P_o$).
(j) Flue gas flow is 1000 scfm ($F_f$).
(k) Atmospheric pressure is 408 inches w.c. ($P_a$).
(l) Disturbance is a rapid 20 percent increase in inlet pressure ($\Delta L$).
(m) Inlet pressure (discharge of forced draft fan) is 15 inches w.c. ($P_i$).

(New information)

(n) Pressure transmitter is Rosemount 1151DP with maximum time constant. ($TC_m$ = 1.7 seconds)

*Find:* the new peak and accumulated errors
*Solution:*
(a) Calculate the new loop period:

$$T_u = 4*\left[1 + \left[\frac{TC}{TD}\right]^{0.65}\right]*TD \tag{4.15}$$

$$TC1_1 = TC1 \text{ (}TC1\text{ from example 4.64)}$$

$$TC1_2 = TC2 \text{ (}TC2\text{ from example 4.64)}$$

$$TC2 = TC_m \tag{4.2}$$

$$TD = Y_1*TC1_1 + Y_2*TC1_2$$

$$TD = 0.5*0.5 + 0.4*1.2 = 0.7 \text{ second}$$

$$T_u = 4*\left[1 + \left[\frac{1.7}{0.7}\right]^{0.65}\right]*0.7$$

$$= 8 \text{ seconds}$$

(b) Calculate the new peak and accumulated errors:

$$PB = \frac{K_g*100*T_u}{2\pi} \tag{4.17}$$

$$PB = \frac{1.5*100*8}{2\pi} = 190 \text{ percent}$$

$$E_x = \left[\frac{1.1*PB}{100}\right]*E_o \tag{3.11}$$

$$E_x = \left[\frac{1.1*190}{100}\right]*0.29$$

$$= 0.6 \text{ inch w./c.}$$

(The closed loop error can be larger than the open loop error for an integrating process since the open loop error is a rate of change.)

$$E_i = \left[\frac{E_x}{1.1}\right]*0.5*T_u \qquad \text{(Equation 4.20 substituted into 4.19)}$$

$$E_i = \left[\frac{0.6}{1.1}\right]*0.5*8 = 2.2 \text{ (inches w.c.)*seconds}$$

*Conclusions:* The loop period increased by a factor of 3. The peak error increased by a factor of 6 and the accumulated error increased by a factor of 12. The use of a smaller time constant is typically not possible because of significant measurement noise. The predicted peak and accumulated errors are still significantly less than those experienced in the field because the effects of slow dampers, measurement noise, and interaction have not been included.

### 6.4.5 EXOTHERMIC REACTOR TEMPERATURE LOOP (RUNAWAY)

*Given:*

(Information from Example 4.6.5)

(a) Set point is 150 F.

(b) Measurement range is 100 to 200 F.
(c) Reactant feed flow is 2000 pph (*Wr*).
(d) Reactant mass is 3000 pounds (*Mr*).
(e) Reactant heat capacity is 0.5 Btu/(lb*F) ($C_r$).
(f) Heat transfer coefficient*area is 8000 Btu/(hr*F) (*UA*).
(g) Change in heat generation with temperature is 12,000 Btu/F ($\Delta Q/\Delta T$).
(h) Coolant mass is 400 pounds ($M_c$).
(i) Coolant heat capacity is 1 Btu/(lb*F) ($C_c$).
(j) Coolant flow is 80,000 pph ($W_c$).
(k) Disturbance is a 20 percent increase in coolant temperature ($\Delta L$).
(l) Coolant temperature is 100 F.
(m) Continuous reactor.

(New information)

(n) Temperature sensor is thermocouple in thermowell.
(o) Annular clearance in thermowell is 0.01 inch.
(p) Annular fill in thermowell is air.
(q) Liquid velocity at thermowell is 1 fps.
(r) Chromatograph sample time is 30 minutes ($T_s$).

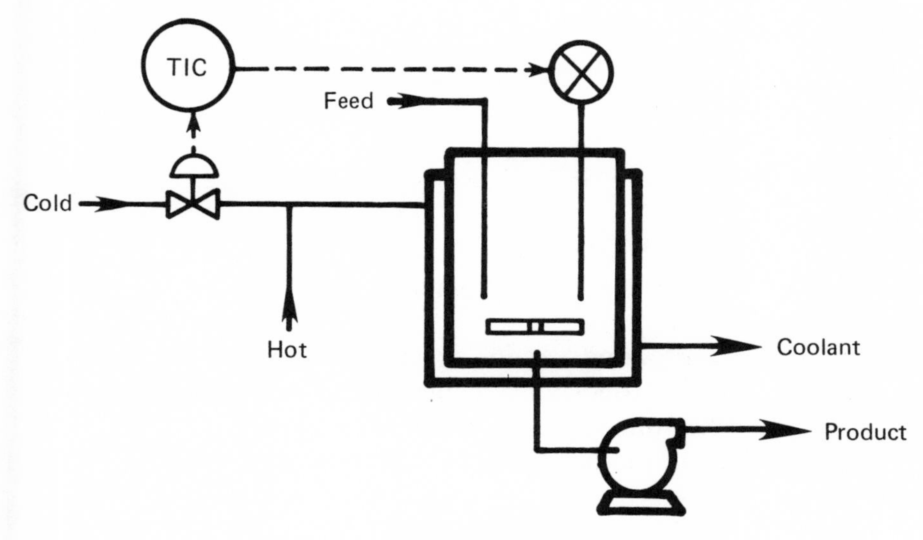


FIGURE 6.8
Exothermic reactor temperature loop.

*Find:* the new peak and accumulated errors for a temperature sensor and alternately for a chromatograph.

*Solution:*

(a) Calculate the new ultimate period for the temperature sensor:

$$TD = TD_p + Y{*}TC_{m1} + Y{*}TC_{m2}$$
$$TD = 0.3 + 0.6{*}0.43 + 0.9{*}0.067$$
$$= 0.62 \text{ minute}$$

(See Table 6.2 for the thermocouple and thermowell time constants.)

$$T_u = 4{*}\left[1 + \left[\frac{N}{D}\right]^{0.65}\right]{*}TD$$

$$N = 2970 \text{ from Example 4.4.5}$$

$$D = (TC' - TC){*}(TC' - TD){*}TD$$

$$D = (30 - 3){*}(30 - 0.62){*}0.62 \tag{4.21}$$

$$= 492$$

$$T_u = 4{*}\left[1 + \left[\frac{2970}{492}\right]^{0.65}\right]{*}0.62$$

$$= 10.5 \text{ minutes}$$

For the chromatograph:

$$TD = TD_p + 1.5{*}T_s$$
$$TD = 0.3 + 1.5{*}30 = 45.3 \text{ minutes}$$

An ultimate period cannot be calculated since the loop is unstable for all proportional bands because the dead time is larger than the runaway time constant ($TD > TC'$).

(b) Calculate the new peak and accumulated errors:

$$E_x = \left[\frac{1.1{*}K_g{*}T_u^{\,2}}{(2\pi)^2{*}TC{*}TC'}\right]{*}E_o \tag{4.25}$$

$$E_x = \left[\frac{1.1*1.5*10.5^2}{(2\pi)^2*3*30}\right]*53$$

$$= 2°\text{F}$$

$$E_i = \left[\frac{E_x}{1.1}\right]*0.5*T_u \qquad \text{(Equation 4.25 substituted into 4.24)}$$

$$E_i = \left[\frac{2}{1.1}\right]*0.5*10.5 = 9.7°*\text{minutes}$$

*Conclusions:* The ultimate period increased by factor of 1.4 for the temperature sensor. The peak error also increased by a factor of 1.4 and the accumulated error increased by a factor of 2 for the temperature sensor. The multiplier for the dead time in the equation for the ultimate period decreased because the runaway time constant is much larger than the dead time in the denominator term. As the dead time approaches the runaway time constant in magnitude, the denominator approaches zero and the multiplier approaches infinity. Thus the effect of dead time is extremely nonlinear. The chromatograph cannot be used for closed loop control because the dead time due to sampling is greater than the runaway time constant.

### 6.4.6 BIOLOGICAL REACTOR CONCENTRATION LOOP (RUNAWAY)

*Given:*

(Information from Example 4.6.6)

(a) Average nutrient concentration is 4 ng/ml ($C_i$).
(b) Nutrient concentration where growth rate is half its maximum is 1 ng/ml ($K_i$).
(c) Maximum cell growth rate is 0.01 generation per minute ($U_x$).
(d) Substrate consumption rate for cell survival is 0.001 generation per minute ($K_e$).

(e) Reactor working volume is 600 gallons ($V$).
(f) Average throughput flow is 10 gpm ($F$).
(g) Agitator pumping rate is 200 gpm ($F_a$).
(e) Loop steady-state gain is 1 ($Kv*K_p*K_m$)

(New information)

(f) Turbidity meter sample transportation delay is 2 minutes.

*Find:* the new *PB* window size.
*Solution:*
(a) Calculate the new ultimate period for the analysis dead time:

$$Tu = 4*\left[1 + \left[\frac{N}{D}\right]^{0.65}\right]*TD$$

$$N = (TC' + TC)*TC'*TC = 1{,}610{,}706 \text{ from Example 4.6.6)} \quad (4.21)$$

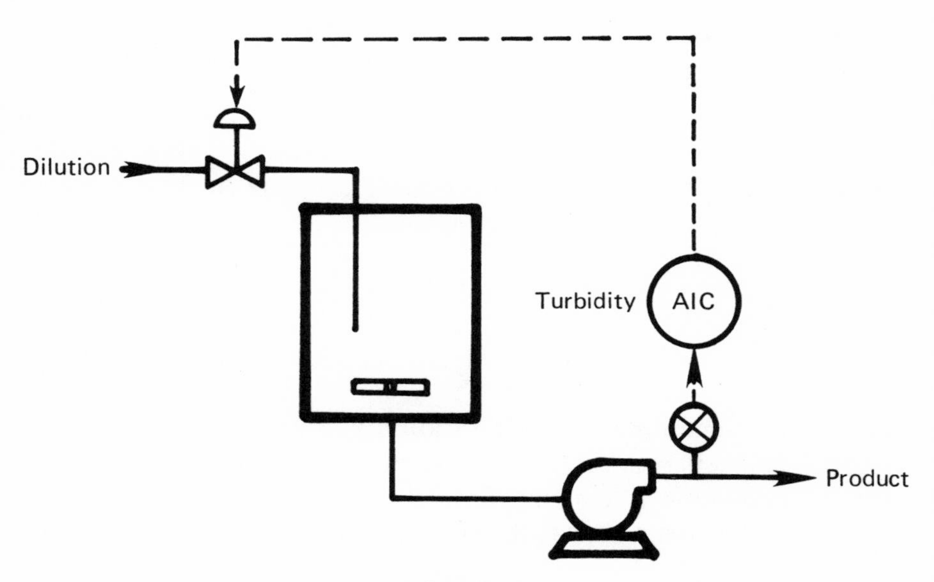


FIGURE 6.9
Biological reactor concentration loop.

$$D = (TC' - TC)*(TC' - TD)*TD$$

$$D = (142 - 57)*(142 - 5)*5 = 58{,}225$$

$$Tu = 4*\left[1 + \left[\frac{1{,}610{,}706}{58{,}225}\right]^{0.65}\right]*5 = 193 \text{ minutes}$$

(b) Calculate the minimum and maximum proportional bands:

$$PB_{\text{max}} = 80*K_v*K_p*K_m = 80 \text{ percent} \tag{4.26}$$

$$PB_{\text{min}} = 0.5*PB \tag{4.27}$$

$$PB = \frac{K_g*100*Tu^2*K_v*K_p*K_m}{(2\pi)^2*TC*TC'} = \frac{5{,}592{,}504}{319{,}214} \tag{4.22}$$

$$= 18 \text{ percent}$$

$$PB_{\text{min}} = 0.5*18 = 9 \text{ percent}$$

(c) Calculate the $PB$ window size:

$$\frac{PB_{\text{max}}}{PB_{\text{min}}} = \frac{80}{9} = 8.9$$

*Conclusions:* The window size decreased significantly. Sample transportation delays must be minimized.

# CHAPTER VII
# Effect of Valve Dynamics

## 7.1
## Valve Time Constant

All time constants that exist between the controller output signal and the disturbance location can be classified as valve time constants. The time constants of valve positioners, valve I/P transducers, and valve boosters are negligible. The dynamic response of the control valve and actuator is a rate-limited exponential. The maximum rate of change of valve position is limited by the stroking speed of the valve and actuator combination. Figure 7.1 shows the rate-limited open loop response of a control valve for a fixed stroking time and variable inherent time constant to a step input signal. The effective time constant depends on the stroking speed, an inherent time constant, and the stroke size. The effective time constant can be estimated by Equation 7.1 where the inherent time constant is $Rv*Tv$ ($Tv$ is the time for a full-scale stroke, which is the inverse of the stroking speed). Equation 7.1 approximates the graphical technique shown in Figure 4.1.

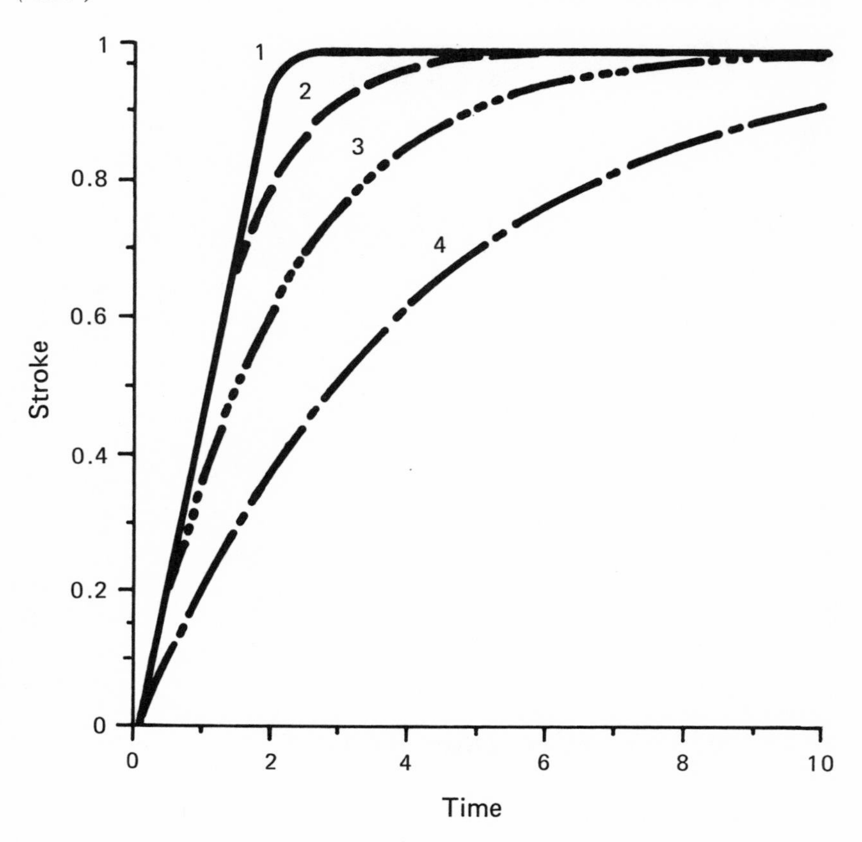


Legend

1 $R_v$ = 0.1

2 $R_v$ = 0.5

3 $R_v$ = 1.0

4 $R_v$ = 2.0

Effect of 0.1 to 2.0 Time-constant to
Stroking-time ratio on step signal

FIGURE 7.1
Rate-limited open loop response.

$$TC_v = T_v*(\Delta V + R_v) \qquad (7.1)$$

where

$TC_v$ = effective time constant of the control valve and actuator
$T_v$ = time for a full-scale stroke
$\Delta V$ = fractional change in valve position
$R_v$ = ratio of inherent time constant to stroking time (for example: $R_v = 0.30$ for piston and $R_v = 0.03$ for diaphragm actuators)

Equation 7.1 shows that the effective time constant is approximately equal to the inherent time constant, $R_v*T_v$, for small changes in valve position and is approximately equal to $T_v$ for full-scale changes in valve position. The stroking time, $T_v$, can be estimated by the use of Equations 7.2 and 7.3 and data from Table 7.1. Equation 7.3 is used to calculate the effective total flow coefficient, $C_v$, due to a restriction such as solenoid valve on the output of the signal source directly supplying the actuator (I/P, positioner, or booster).

$$T_v = \frac{Y_v}{C_v} \qquad (7.2)$$

For a restriction in series with the source:

$$C_v = \left[\frac{(C_{v1}^2)*(C_{v2}^2)}{(C_{v1}^2) + (C_{v2}^2)}\right]^{0.5} \qquad (7.3)$$

where

$T_v$ = valve stroking time (time for a full-scale stroke) in seconds
$Y_v$ = valve stroking time factor from Table 7.1
$C_v$ = effective total $C_v$ (flow coefficient) of the valve accessories
$C_{v1}$ = $C_v$ (flow coefficient) of valve accessory no. 1
$C_{v2}$ = $C_v$ (flow coefficient) of valve accessory no. 2

TABLE 7.1
Control Valve Prestroke Dead Time and Stroking Time Factors

| Actuator, sq. in. | Travel, inches | Pressure, psig | Spring, lb/in. | Fill Factors $X_v$ | Fill Factors $Y_v$ | Exhaust Factors $X_v$ | Exhaust Factors $Y_v$ |
|---|---|---|---|---|---|---|---|
| | | | *Diaphragm* | | | | |
| 66 | 21/8 | 4.5–16 | 275 | 0.015 | 0.338 | 0.225 | 0.610 |
| 100 | 31/2 | 6–22 | 335 | 0.031 | 0.861 | 0.404 | 1.446 |
| 215 | 41/8 | 6–21 | 610 | 0.105 | 2.276 | 1.200 | 3.902 |
| 46 | 3/4 | 3–15 | 735 | 0.012 | 0.190 | 0.045 | 0.256 |
| 46 | 3/4 | 6–30 | 1470 | 0.016 | 0.226 | 0.031 | 0.290 |
| 69 | 3/4 | 3–15 | 1100 | 0.020 | 0.297 | 0.071 | 0.401 |
| 69 | 3/4 | 6–30 | 2210 | 0.028 | 0.355 | 0.048 | 0.457 |
| 105 | 3/4 | 3–15 | 1670 | 0.033 | 0.466 | 0.115 | 0.630 |
| 105 | 3/4 | 6–30 | 3320 | 0.046 | 0.574 | 0.078 | 0.727 |
| 156 | 3/4 | 3–15 | 2500 | 0.046 | 0.676 | 0.161 | 0.913 |
| 156 | 3/4 | 6–30 | 5000 | 0.065 | 0.811 | 0.110 | 1.046 |
| 220 | 2 | 3–15 | 1260 | 0.074 | 2.004 | 0.500 | 2.810 |
| 220 | 2 | 6–30 | 2520 | 0.104 | 2.243 | 0.390 | 3.038 |
| 310 | 2 | 3–15 | 1650 | 0.103 | 2.790 | 0.569 | 3.724 |
| 310 | 2 | 6–30 | 3100 | 0.143 | 3.379 | 0.388 | 4.265 |
| 450 | 2 | 6–26 | 4500 | 0.323 | 4.386 | 1.260 | 6.552 |
| 450 | 2 | 6–26 | 4500 | 0.380 | 4.586 | 1.353 | 6.870 |
| | | | *Piston* | | | | |
| 17 | 3/4 | 60 | — | 0.085 | 0.050 | 0.024 | 0.050 |
| 28 | 3/4 | 60 | — | 0.165 | 0.086 | 0.035 | 0.086 |
| 56 | 3/4 | 60 | — | 0.296 | 0.169 | 0.050 | 0.169 |
| 89 | 2 | 60 | — | 0.715 | 0.719 | 0.196 | 0.719 |
| 131 | 2 | 60 | — | 0.995 | 1.060 | 0.272 | 1.060 |
| 222 | 2 | 60 | — | 1.730 | 1.800 | 0.738 | 1.800 |
| 17 | 4 | 60 | — | 0.020 | 0.278 | 0.024 | 0.278 |
| 28 | 4 | 60 | — | 0.051 | 0.460 | 0.035 | 0.460 |
| 56 | 4 | 60 | — | 0.099 | 0.901 | 0.050 | 0.901 |
| 89 | 4 | 60 | — | 0.181 | 1.453 | 0.196 | 1.453 |
| 131 | 4 | 60 | — | 0.227 | 2.144 | 0.272 | 2.144 |
| 222 | 4 | 60 | — | 0.603 | 3.600 | 0.738 | 3.600 |

| Accessory Type | Connection Sizes, inches | Supply $C_v$ | Exhaust $C_v$ |
|---|---|---|---|
| Positioner | — | 0.17 | 0.24 |
| I/P transducer | — | 0.39 | 0.36 |
| Quick release valve | 3/8 NPT | 2.94 | 3.6 |
| Quick release valve | 1/2 NPT | 5.46 | 6.49 |
| Quick release valve | 3/4 NPT | 9.97 | 12.17 |
| Volume booster | 3/8 and 3/8 ports | 3.74 | 2.29 |
| Volume booster | 3/8 and 1/2 ports | 3.74 | 2.52 |
| Volume booster | 1/2 and 3/8 ports | 5.32 | 2.30 |
| Volume booster | 1/2 and 1/2 ports | 5.32 | 2.53 |
| Volume booster* | 1 1/2 NPT | 23.1 | 23.1 |

* Fairchild 200 XLR

If the control valve time constant is not the largest in the loop, the control valve time constant should be converted to equivalent dead time via Equations 4.1 and 4.2 ($TC1 = TC_v$) and added to the total loop dead time for use in the equations in Chapter 4. If the control valve time constant is the largest time constant in the loop, all the other time constants in the loop should be converted to equivalent dead time via Equations 4.1 and 4.2 ($TC2 = TC_v$) added to any pure dead times for use in the equations in Chapter 4. If the control valve time constant is the largest time constant in the loop, the control valve time constant should be used as the loop time constant (*TC*) in the equations in Chapter 4 and the peak and accumulated errors should be multiplied by the ratio of the control valve to process time constant. These then are the errors in the controlled variable.

If $TC_v$ becomes larger than 5 percent of the dead time and the controller mode settings are not readjusted, its effect on loop period can be estimated by Equation 6.1 for self-regulating processes and Equation 6.3 for integrating processes (substitute $TC_v$ for $TC_m$ in the equations). The time duration of the first quarter cycle of the closed loop response is about one fourth of the ultimate period of the loop with a negligible valve time constant, since the valve does not start to respond appreciably until after the first quarter cycle (see Figure 7.2).

If the loop has a large measurement time constant as well as a large control valve time constant, then the increase in loop period due to the slow measurement should be multiplied by the increase in loop period due to the slow valve to estimate the change in loop period for self-regulating processes. The effects are multiplicative for self-regulating and additive for integrating processes.

Any equipment or process time constant between the controller output and the disturbance that slows down the response of the manipulated variable has the same effect as a slow valve. The effect of a solid reagent solubility time constant (e.g., lime) on pH control loop performance can be estimated by the use of Equation 6.1 for continuous operation and Equation 6.3 for batch operation. (The lime may also be coated and settled out of solution by a precipitate such as $CaSO_4$.)

The most common mistake made during the design of a surge

control valve installation for vent or recycle flow is to ignore the need to throttle quickly in either direction. Frequently the valve accessories are installed to provide a fast, full-scale stroke of 1 second or less but require 5 seconds or more to throttle to intermediate positions. The use of used quick exhaust valves, small-orifice-solenoid valves, small-flow-capacity pilots, and small-diameter tubing for accessories on eccentric disk valves

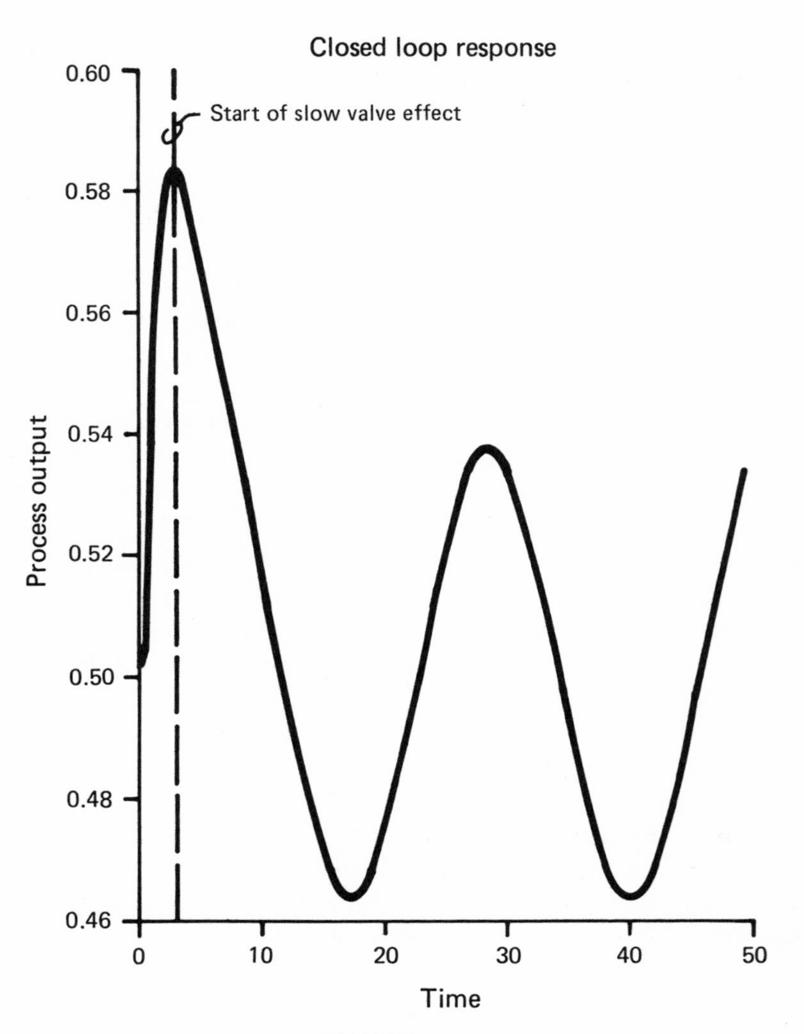


FIGURE 7.2
Closed loop response for a large valve time constant ($TC_v$ = 100*$TD$ for a self-regulating process).

results in a large overshoot near the closed position, slow throttling, and erratic stroking for inadvertent triggering of the quick exhaust valve (see Figure 7.4 for a plot of the nonlinear torque requirement of the eccentric disk valve that causes the overshoot problem). Figure 7.3 shows the proper installation of accessories to ensure precise and fast throttling in addition to full-scale fast stroking for a diaphragm actuator. These same accessories are applicable to furnace pressure control dampers or to any control valve in a critical fast loop. Piston actuators should be avoided because two boosters are required and the combination of a booster and positioner in series is unstable unless the booster is detuned by bypassing some of the air signal around the booster (Mamzic, 1958). Some boosters have integral bypasses and an adjustment screw. The pneumatic connections are similar to that shown in Figure 7.3, except that the booster's input signal is the positioner's output signal and the positioner's input signal is the I/P's output signal after passing through the interlock solenoid valve.

The use of positioners on surge valves should be avoided. A positioner slows down the valve response because its flow capacity is typically small (the $C_v$ of a positioner is less than that of an I/P). A positioner will require detuning not only of the booster, but also of the surge controller because the surge loop is a "fast" loop. The use of a positioner on a fast loop to solve a hysteresis problem will only make the problem worse (Lloyd, 1969). A positioner creates a cascade loop where the "inner" valve position loop is slower than the "outer" surge loop (see Chapter 11).

An electronic position transmitter is recommended on critical fast valves because valve speed deteriorates with age, an im-

TABLE 7.2
Control Valve Hysteresis

| *Valve Type* | *Packing Type* | *Positioner* | *Pressure Range, psig* | *Hysteresis, %* |
|---|---|---|---|---|
| Sliding stem | TFE | No | 3–15 | 5 |
| Sliding stem | TFE | No | 6–30 | 5 |
| Sliding stem | TFE asbestos | No | 3–15 | 10 |
| Sliding stem | TFE asbestos | No | 6–30 | 7 |
| Sliding stem | Laminated graphite | No | 3–15 | 15 |
| Sliding stem | Laminated graphite | No | 6–30 | 10 |
| Sliding stem | — | Yes | — | 1 |

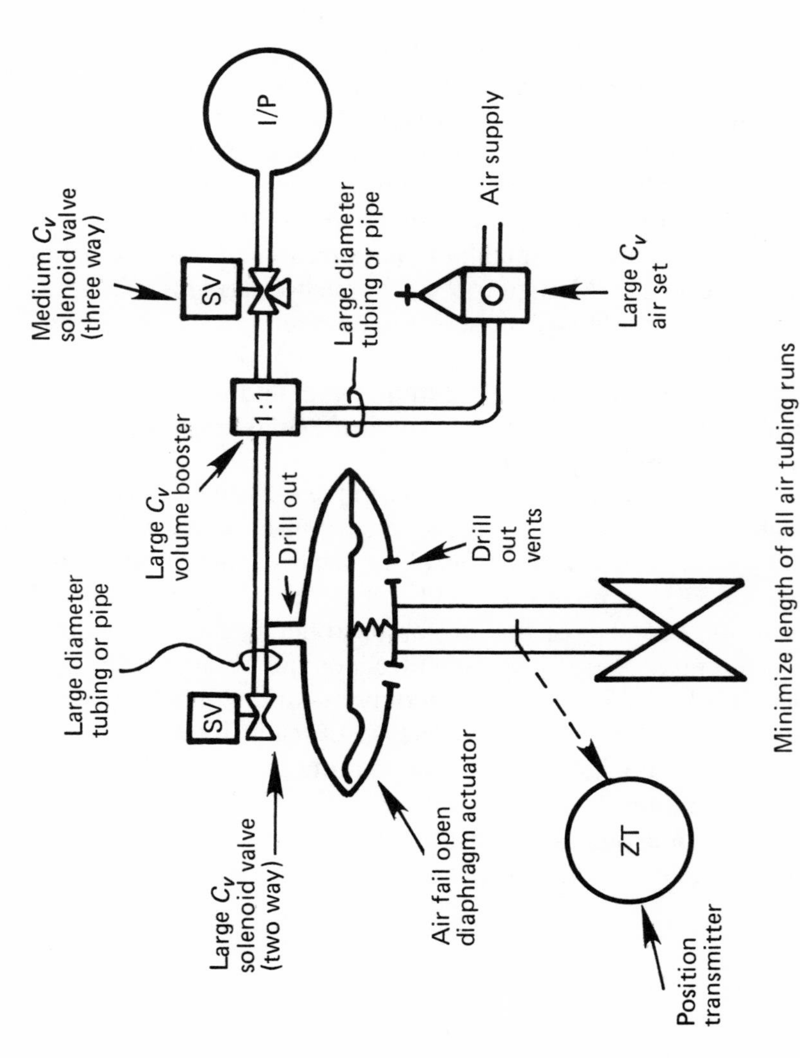


FIGURE 7.3
Surge control valve accessories.

properly tuned booster can cause erratic stroking, and the stroking time is too fast to be monitered visually directly. The position transmitter output should be an input to either a high-speed recorder or a data logger. The prestroke dead time and stroking time are too small to check visually with a stopwatch.

## 7.2 Valve Dead Time

Valve dead time can originate from multiple time constants in series or prestroke dead time. Since the time constants of valve accessories are usually negligible, valve dead time due to multiple time constants is important only if there are process or equipment time constants before the entry point of the disturbance in the loop. Valve prestroke dead time occurs when the valve is stationary or needs to reverse direction. The valve dead time can be estimated by the use of Equations 7.3 and 7.4 and data from Table 7.1.

$$TD_v = \frac{X_v}{C_v} \tag{7.4}$$

where

$TD_v$ = valve dead time
$X_v$ = prestroke dead-time factor from Table 7.1
$C_v$ = effective total $C_v$ (flow coefficient) of the valve accessories

The prestroke dead time strongly depends on the direction of the stroke and can be larger than the stroking time (see Table 7.1). The prestroke dead time that is estimated from the data in Table 7.1 does not take into account valve hysteresis due to stroke friction. Rotary valves, grafoil packing, stem corrosion, and stem coatings cause hysteresis, which should be included by increasing the prestroke dead time by an amount equal to the valve stroking time multiplied by the hysteresis expressed as a fraction of the full-scale stroke (see Table 7.2).

If the stroke travel of interest is not listed, the same $X_v$ can be used but the $Y_v$ should be changed in proportion to the change in travel. The filling time corresponds to the upstroke time and the exhaust time corresponds to the downstroke time for piston actuators.

The booster port sizes are for the supply and exhaust ports respectively. Caution must be exercised in applying a booster and quick release valve in series because the rapid pressure changes initiated by the booster can inadvertently trigger the quick release valve.

## *7.3 Valve Stroke Accuracy and Rangeability*

Valve stroke accuracy is limited principally by valve hysteresis. Valve hysteresis is the difference in valve stroke between an increasing and decreasing valve signal expressed as a percent of the valve stroke. Hysteresis includes both hysteretic error and dead band. It is an indicator of the change in valve signal required to change the direction of the stroke or to start the stroke from a stationary position (see Figure 7.5). Valve stroke resolution is approximately equal to valve hysteresis for a change in direction of stroke or to valve dead band for the start of the stroke. If the valve is continuing to stroke in one direction, the valve stroke resolution is smaller. The valve stroke resolution for an eccentric disk valve opening may be as poor as 30 percent because the large breakaway torque followed by the large dip in torque requirement causes the valve to overshoot to 30 percent from the closed position. The overshoot can be reduced but not eliminated by the addition of volume boosters. The overshoot is particularly troublesome for closed loop compressor control. Posi-Seal International, Inc., has developed a mechanical linkage to compensate automatically for the nonlinear torque requirement of its eccentric disk valve.

Control valve hysteresis can be reduced on slow loops by the addition of a valve positioner. On fast loops, the use of a positioner may actually increase the probability of a limit cycle because of hysteresis. If the valve response is slower than the process response, the controller will have to be detuned (the proportional band increased) regardless of the degree of hysteresis because the positioner is the inner loop of a cascade loop (see Chapter 11). Hysteretic and dead-band effects are particularly detrimental for integrating and runaway processes. Describing functions and Nyquist plots can be used to predict limit cycling in such non-self-regulating processes (Brez, 1975). The high measurement gain of pH loops amplifies the effect of valve hysteresis. A 1 percent valve hysteresis in a strong acid and strong base neutralization can cause a 3 pH offset from set point (McMillan, 1981a). The increased popularity of rotary valves has increased the incidence of loop performance problems due to valve hysteresis. Tests on hysteresis at the factory are made with the packing-only hand tight. The hysteresis of rotary valves with TFE packing is about the same as the hysteresis of sliding stem valves with grafoil packing. Since the

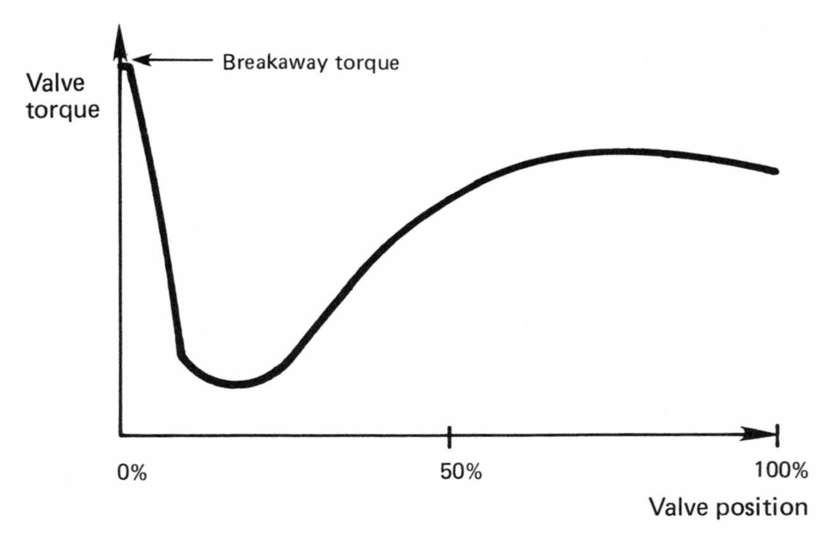


FIGURE 7.4
Nonlinear torque requirement of an eccentric disk valve.

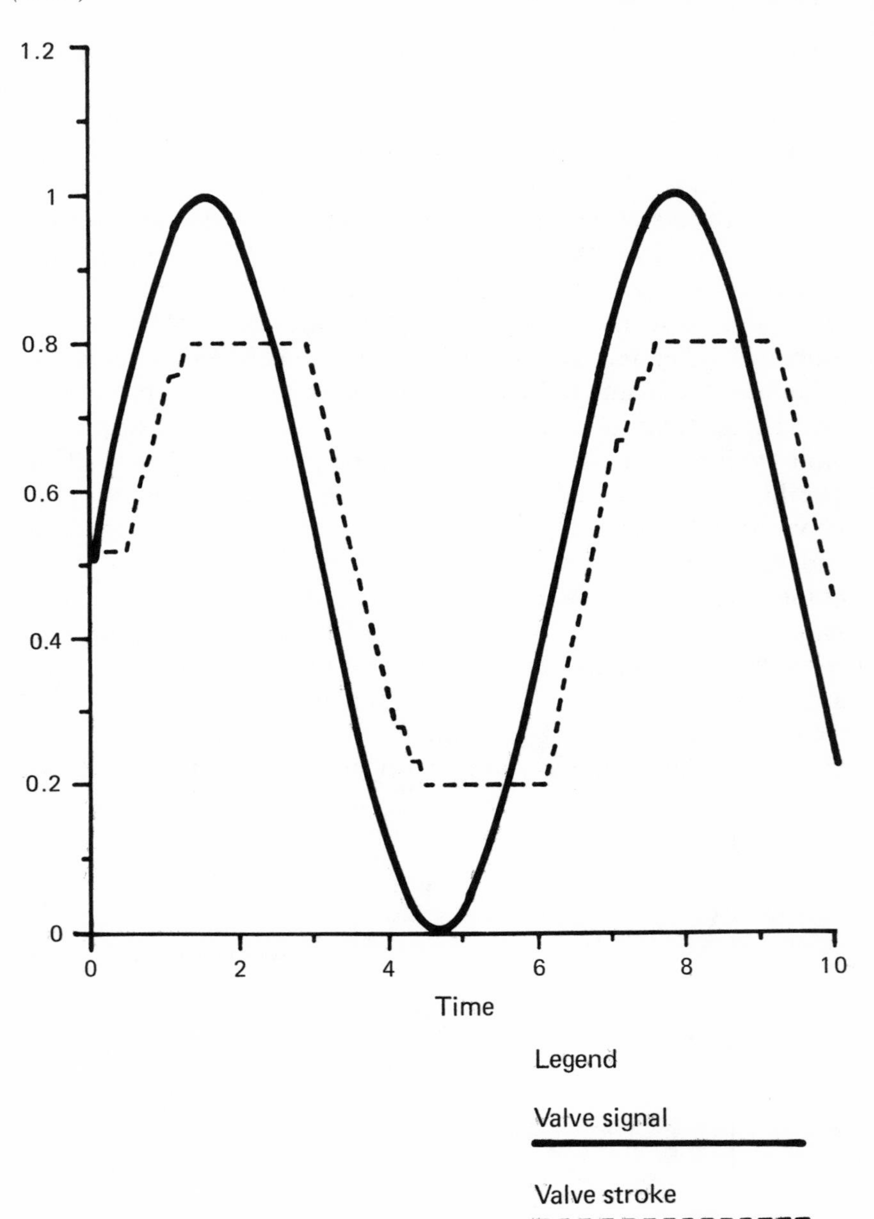


FIGURE 7.5
Effect of control valve hysteresis on loop oscillation.

positioners frequently used with rotary valves are not high performance (high gain), the hysteresis is not reduced to 1 percent by the addition of the positioner.

Valve inherent rangeability is the ratio of the maximum to the minimum controllable flow coefficient within which the inherent flow characteristic does not exceed some specified limits. The inherent flow characteristic is the relationship between flow and travel as the travel is varied from 0 to 100 percent for a fixed inlet pressure and pressure drop. These definitions are as stated in Schuder (1971), except for the addition of "controllable" and "inlet pressure." It is advisable to include the word "controllable" because valve stroke resolution places a practical limit on the minimum controllable flow coefficient. The word "inlet pressure" should be included because gas flow also depends on the inlet pressure. If the effect of valve stroke resolution is ignored, the quick opening inherent flow characteristic yields the best—and the linear inherent flow characteristic the next best—inherent rangeability. However, if the effect of valve stroke resolution is included and the valve stroke resolution at shutoff is worse than 2 percent, the opposite is true. The equal percentage inherent flow characteristic gives a 50:1 inherent rangeability for a wide range of valve stroke resolutions due to the flatness of the characteristic curve at shutoff. The literature has ignored the effect of valve stroke resolution on valve rangeability specifically, and on loop performance in general. Since rangeability in reality depends on resolution and resolution depends on hysteresis, meeting rangeability or resolution requirements means meeting a hysteresis requirement.

Valve installed rangeability depends on the variation in pressure drop for liquid systems and also on inlet pressure for gas systems. The installed rangeability can be estimated by Equa-

TABLE 7.3
Valve Inherent Rangeability

| *Inherent Characteristic* | *Stroke Resolution* 1% | 10% | 20% |
|---|---|---|---|
| Quick opening | 50:1 | 6:1 | 3:1 |
| Linear | 100:1 | 10:1 | 5:1 |
| Equal percentage | 50:1 | 50:1 | 40:1 |

tion 7.5 for liquid systems, if the inherent rangeability and the pressure drops at maximum and minimum flow are known (Shinskey, 1978).

$$R_f = R_i{*}\left[\frac{\Delta P1}{\Delta P2}\right]^{0.5} \tag{7.5}$$

where

$R_f$ = installed (final) rangeability
$R_i$ = inherent rangeability
$\Delta P1$ = pressure drop at maximum flow
$\Delta P2$ = pressure drop at minimum flow

For liquid pump and valve systems, $\Delta P1$ is less than $\Delta P2$ since the pump head curve decreases with flow and the system frictional drop increases with flow. The resulting installed rangeability may not be large enough to allow control at the minimum flow. The installed flow characteristic of a linear inherent characteristic distorts toward a quick opening characteristic and the installed characteristic of an equal percentage inherent flow characteristic distorts toward a linear characteristic as the ratio $\Delta P1/\Delta P2$ decreases. The popularity of the equal percentage characteristic is due to the nearly linear valve gain of its installed characteristic and to the greater inherent rangeability for poor stroke resolution at shutoff.

The rangeability for a loop can be increased by split ranging multiple control valves. However there is a severe discontinuity in the valve gain at the transition from one valve to another and the valve resolution problem still exists. Pressure switches on the pneumatic signals to sequence the control valves have been used with varying degrees of success. The pressure switch setting resolution and repeatability and the valve prestroke dead time and stroking time may cause problems in some installations. Valve resolution can be increased by the use of multiple valves in parallel, each with its own controller with the proportional band set to match the individual valve gain. However the individual controllers will fight each other if the speed of response of each valve is not identical to the others. The following special valves should be considered for those applications that require high rangeability and high stroke resolution.

### 7.3.1 LAMINAR FLOW VALVE

The laminar flow valve developed by Hans D. Baumann (developer of the Camflex® valve) provides tremendous inherent rangeability for small flow ranges ($C_v < 0.01$) because the flow characteristic varies with the third power of the stroke (see Figure 7.6). For a stroke range of 25:1, the inherent rangeability is 15,000:1 if the stroke resolution is 1 percent or better at shutoff. Since the change in pump head and system frictional pressure drop is negligible for these small flows, the inherent rangeability is the installed rangeability. The control valve gain is proportional to the second power of the stoke. This valve gain is easily compensated for in a microprocessor controller because it is fixed by the laminar flow and constant pressure drop (Baumann, 1981). The transition from turbulent to laminar flow in a conventional control valve causes a large change in the actual flow coefficient, and hence the valve gain and rangeability, because flow is proportional to the pressure drop to the second power for turbulent flow and to the first power for laminar flow.

### 7.3.2 DIGITAL VALVE

The digital valve developed by Process Systems, Inc. (a division of Powell Industries, Inc.) provides an inherent rangeability determined by the number of binary coded ports.

$$R_i = 2^{N_p} \tag{7.6}$$

where

$R_i$ = inherent rangeability
$N_p$ = number of binary coded ports

 

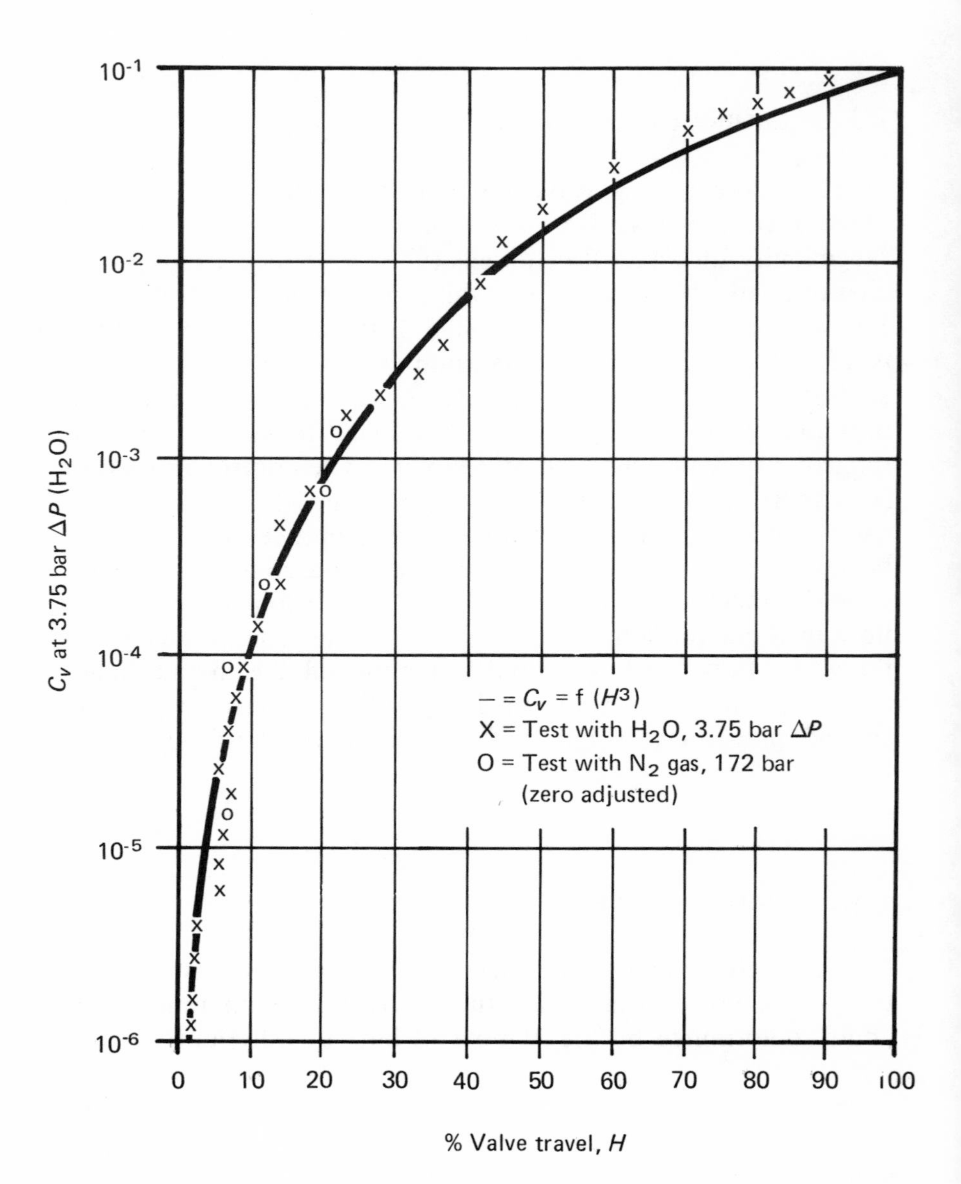


FIGURE 7.6
Measured flow characteristic of laminar flow valve.

The valve stroke resolution and the valve stroke hysteresis are equal to the inverse of the valve rangeability multiplied by 100 percent. Thus the inherent rangeability is 4096:1 and the stroke resolution and hysteresis are both about 0.02 percent for a 12-bit digital valve. The lives of the individual solenoids are prolonged by reducing the power input periodically (the duration of the reduction in power is not long enough to switch the valve), which reduces the operating temperature of the solenoids. The life of the smallest port and solenoid is prolonged by use of a filter on the valve signal to prevent dithering. The inherent flow characteristic is linear, the prestroke dead time and stroking time total varies from 25 to 100 milliseconds, depending on valve size, and the overshoot is zero. These features make the digital valve ideally suited for compressor surge control. There are some important problems to be reckoned with in the maintenance of digital valves. Since the digital valve uses soft seats, it is susceptible to erosion damage by particulates or flashing. The sticking or failure of an individual actuator cannot be determined by external inspection. The smallest part is susceptible to pluggage in fouling service since it is the most frequently used port.

### 7.3.3 PULSE INTERVAL CONTROL

Pulse interval control provides an inherent rangeability approximately equal to the maximum interval divided by the pulse width selected. The maximum interval should be less than twice the residence time of the tank. The pulse width should be just larger than the total of the valve prestroke dead time and stroking time. Pulse interval control is preferable to pulse width control because the shorter duration pulses are more effectively filtered by the capacitance of downstream equipment. Figure 7.7 shows how the pH measurement signal amplitude is much larger for pulse width modulation for an increase in reagent demand by a pH controller on a well-mixed tank. The stroke resolution and stroke hysteresis are only limited by the resolution of the electronic pulse generator (typically a counter integrator) and the repeatability of the control valve stroke. Extremely small equivalent flows can be delivered by the use of large pulse intervals, which eliminate the plugging problem associated with throttling small orifices. Plug or ball valves with on–off operators not only are inexpensive, but minimize plugging problems (Whitey ball valves have been successfully used on lime reagent for pH control). The control valve gain depends on the pulse width and interval and the average flow capacity of the valve during the pulse.

$$F_v = \left[ \frac{T_p}{\frac{(C_x - C)}{C_x} * T_x + T_p} \right] * F_p \tag{7.7}$$

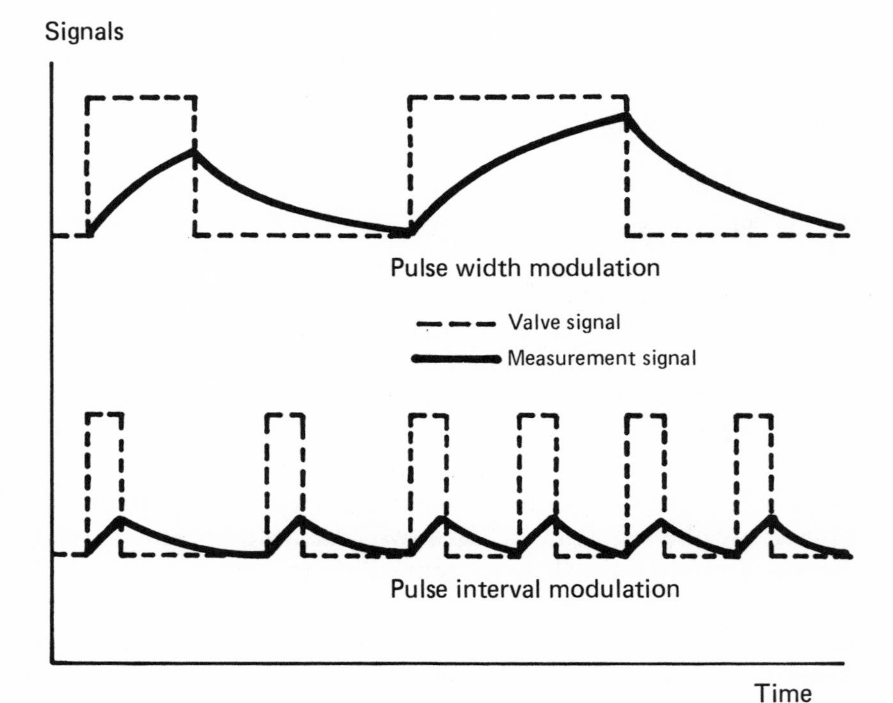


FIGURE 7.7
Comparison of pulse width and pulse interval modulation.

$$K_v = \frac{\Delta F_v}{\Delta C} = \left[ \frac{\dfrac{T_p * T_x}{C_x}}{\left[ \dfrac{(C_x - C)}{C_x} * T_x + T_p \right]^2} \right] * F_p \tag{7.8}$$

For $C = 0$:

$$K_v = \left[ \frac{T_p}{(C_x * T_x)} \right] * F_p \tag{7.9}$$

For $C = C_x$:

$$K_v = \left[ \frac{T_x}{(C_x * T_p)} \right] * F_p \tag{7.10}$$

where

$F_v$ = effective flow through the valve
$K_v$ = control valve gain
$T_p$ = time duration of the pulse
$T_x$ = maximum time interval
$C_x$ = maximum fractional controller output signal
$C$ = fractional controller output signal
$F_p$ = average flow capacity of the valve during the pulse

Since $T_p$ is much less than $T_x$, the valve gain is small for small values of the controller output and large for large values of controller output (see Figure 7.8). Equation 7.8 can be used in a microprocessor controller to provide automatic gain compensation. The inherent characteristic is the installed characteristic because the average flow during the pulse does not change. If the process gain is higher for lower manipulated flows (e.g., constant influent composition and variable influent flow for pH control), then the variable valve gain of pulse interval control may help stabilize the loop. For this case, the ratio of $T_x/T_p$ should equal the ratio of maximum to minimum influent flow. Pulse interval control can potentially save energy for heat exchangers in terms of coolant cost because the valve nonlinearity complements the process nonlinearity due to load changes (less oscillation and therefore less coolant usage) and the cycling of coolant flow increases the heat transfer coefficient. Cyclic flow has also been shown to improve performance of certain gas-solid phase reaction systems and distillation columns (McMillan, 1981b).

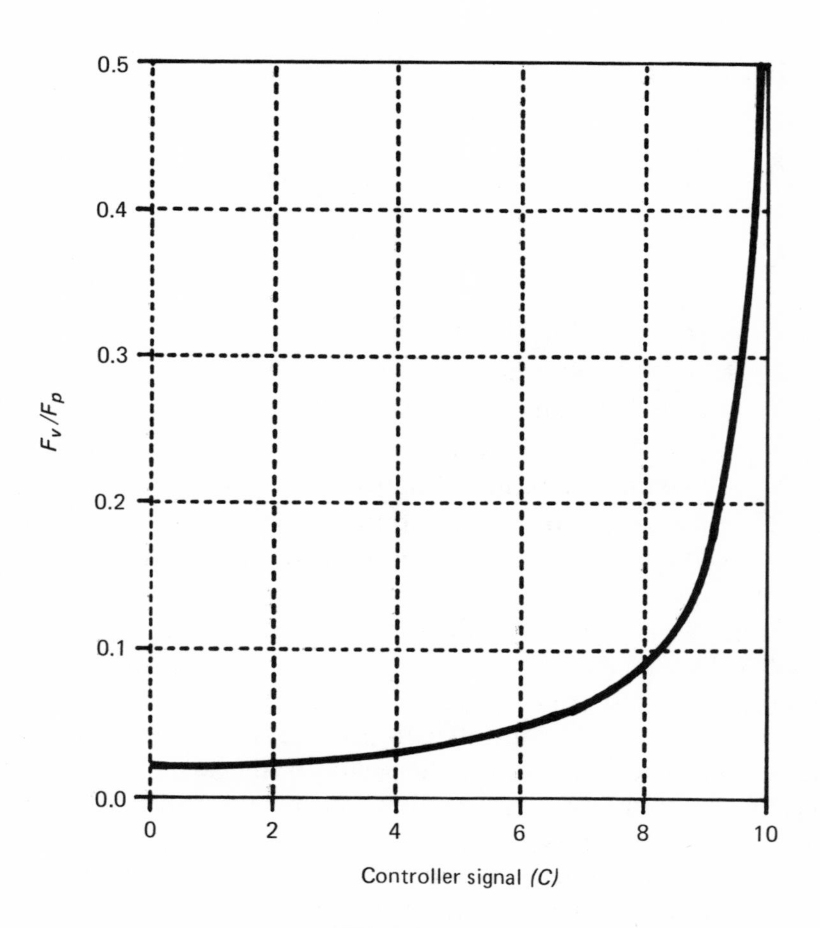


FIGURE 7.8
Flow characteristic of pulse interval control.

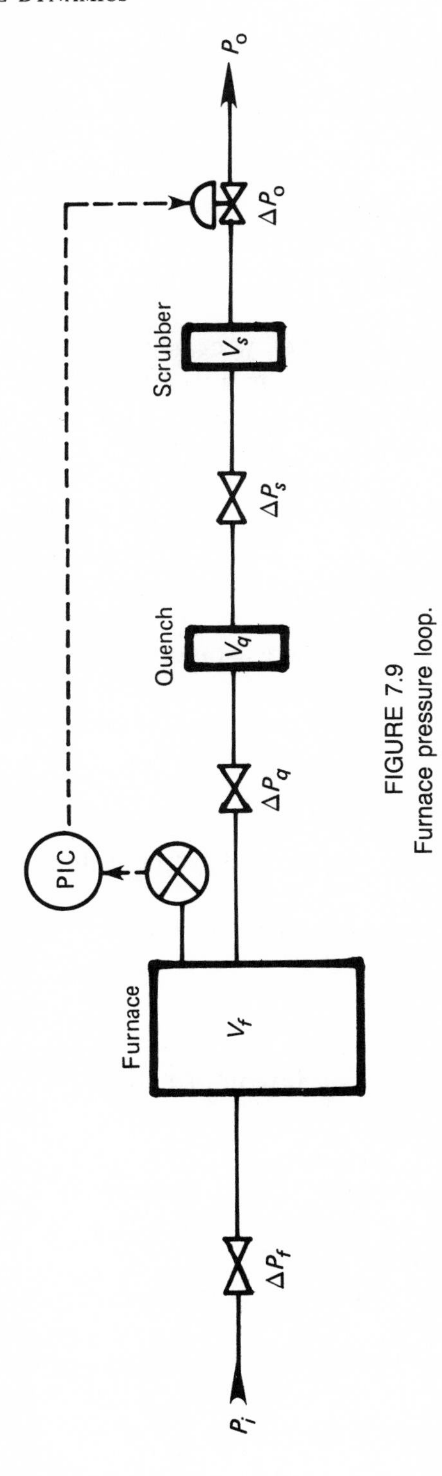


FIGURE 7.9
Furnace pressure loop.

# 7.4 Furnace Pressure Loop Example

*Given:*

(Information from Examples 4.6.4 and 6.4.4)

(a) Set point is 5 inches w.c. gage.
(b) Measurement range is 0 to 10 inches w.c. gage.
(c) Furnace volume is 10,000 ft$^3$ ($V_f$).
(d) Quench volume is 1000 ft$^3$ ($V_q$).
(e) Scrubber volume is 1000 ft$^3$ ($V_s$).
(f) Furnace inlet flow resistance pressure drop is 2.5 inches w.c. ($\Delta P_f$).
(g) Quench inlet flow resistance pressure drop is 5 inches w.c. ($\Delta P_q$).
(h) Scrubber inlet flow resistance pressure drop is 10 inches w.c. ($\Delta P_s$).
(i) System outlet flow resistance pressure drop is 2.5 inches w.c. ($\Delta P_o$).
(j) Flue gas flow is 1000 scfm ($F_f$).
(k) Atmospheric pressure is 408 inches w.c. ($P_a$).
(l) Disturbance is a rapid 20 percent increase in inlet pressure ($\Delta L$).
(m) Inlet pressure (discharge of forced draft fan) is 15 inches w.c. ($P_i$).

(New information)

(n) Damper actuator $Y_v = 1.494$ and $X_v = 0.75$.
(o) Solenoid on positioner outlet $C_v = 0.39$.
(p) Positioner $C_v = 0.17$.
(q) Fractional stroke travel for disturbance is 0.2.

*Find:* the new peak and accumulated errors.
*Solution:*
(a) Find the new loop period:

$$C_v = \left[\frac{C_{v1}^2 * C_{v2}^2}{C_{v1}^2 + C_{v2}^2}\right]^{0.5} \quad (7.3)$$

$$C_v = \left[\frac{0.39^2 * 0.17^2}{0.39^2 + 0.17^2}\right]^{0.5}$$

$$C_v = 0.156$$

$$T_v = \frac{Y_v}{C_v} \quad (7.2)$$

$$T_v = \frac{1.494}{0.156} = 9.6 \text{ seconds}$$

$$TC_v = T_v*(\Delta V + 0.025) \quad (7.1)$$

$$TC_v = 9.6*(0.2 + 0.025)$$

$$= 2.2 \text{ seconds}$$

$$TD_v = \frac{X_v}{C_v} \quad (7.4)$$

$$TD_v = \frac{0.075}{0.156} = 0.5 \text{ second}$$

$$T_u = 4*\left[1 + \left[\frac{TC}{TD}\right]^{0.65}\right]*TD \quad (4.15)$$

$$TC1_1 = 0.5$$

$$TC1_2 = 1.2$$

$$TC1_3 = 1.7$$

$$TC2 = TC_v = 2.2$$

$$TD = Y_1* TC1_1 + Y_2* TC1_2 + Y_3*TC1_3 + TD_v$$

$$TD = 0.5*0.5 + 0.4*1.2 + 0.3*1.7 + 0.7$$

$$TD = 1.9 \text{ seconds}$$

$$T_u = 4*\left[1 + \left[\frac{2.2}{1.9}\right]^{0.65}\right]*1.9 = 16 \text{ seconds}$$

(b) Calculate the new peak and accumulated errors:

$$PB = \frac{K_g*100*T_u}{2\pi} \quad (4.17)$$

$$PB = \frac{1.5*100*16}{2\pi} = 382 \text{ percent}$$

$$E_x = \left[\frac{1.1*PB}{100}\right]*E_o \quad (3.11)$$

$$E_x = \left[\frac{1.1*382}{100}\right]*0.29 = 1.2 \text{ inches w.c.}$$

$$E_i = \left[\frac{E_x}{1.1}\right]*0.5*T_u$$

(Equation 4.20 substituted into 4.19)

$$E_i = \left[\frac{1.2}{1.1}\right]*0.5*22$$

$$= 11 \text{ (inches w.c.)*seconds}$$

*Conclusions:* The loop period increased by a factor of 2. The peak error increased by a factor of 2 and the accumulated error increased by a factor of 5. The elimination of the positioner would have increased the $C_v$ and decreased the $T_v$ by 2.5 but would have increased the hysteresis from 1 percent to about 15 percent. The peak and accumulated errors are starting to approach those in the field. The effect of a given noise signal is calculated in Chapter 8.

CHAPTER VIII

# Effect of Disturbance Dynamics

## 8.1 Disturbance Time Constant

The discussion up to this point has assumed that the disturbance is a rapid change so that a step disturbance could be assumed. If the disturbance is slow, the reduction in errors can be estimated by multiplying the previous errors by the correction factor from Equation 8.1 or 8.2.

$$C_d = e^{\frac{(-4*TC_d)}{(T_u)}} \qquad (8.1)$$

For $TC_d << T_u$:

$$C_d = \frac{T_u}{T_u + 4*TC_d} \qquad (8.2)$$

where

$C_d$ = disturbance correction factor
$TC_d$ = disturbance time constant
$T_u$ = loop period (include the effect of slow transmitter and valve)
$e$ = base of the natural logarithm ($e = 2.71828$)

As the disturbance time constant approaches zero, the correction factor approaches unity. As the time constant approaches infinity, the correction factor approaches zero and the loop error approaches the accuracy limit of the measurement and valve stroke. Thus large tanks upstream of control loops help to reduce control loop errors. The time constant for a concentration disturbance is the time constant of the tank (see Appendix E). However, if the upstream tank's level controller is poorly tuned or consists of high- and low-level switches, severe flow disturbances may develop.

The disturbance can also be too slow. If the operating point approaches a surge controller set point too slowly, reset action integrates the error sufficiently to cause the controller output to return to its maximum value and close the surge valve. This would be more likely to occur with those controllers that center the proportional band around the set point because there would be no proportional action at all until the measurement got within one half of a proportional band of set point. A very slow disturbance can result in the surge measurement being at or slightly above the surge set point with the surge valve shut. Since the prestroke dead time is greatest at the closed position of the surge valve, the surge valve may be too slow for any subsequent disturbance. For a linear approach of the surge measurement to the surge set point, the disturbance time (time for error to go from the initial error to zero error) must be less than twice the reset time setting ($T_i$). For example, if the reset setting were ten repeats per minute ($T_i = 6$ seconds per repeat), a disturbance time of 12 seconds or more for a linear approach to set point would keep the surge valve closed until the measurement crossed set point. If the surge set point were at 50 percent, this would correspond to a block valve stroking time of approximately 24 seconds or more for a block valve with a linear installed flow characteristic (an equal percentage characteristic would give a faster approach and a quick opening characteristic would give a slower approach initially to the surge set point).

A surge control valve that stroked faster to open than close would facilitate a slower closing block valve that had an equal percentage characteristic. However the controller proportional band setting would have to be increased and the reset setting decreased for the overall slower valve. Consequently the peak error, accumulated error, and number of surge cycles would increase for a fast disturbance. For a linear block and surge valve:

$$T_v < T_b < 2^*T_i \tag{8.3}$$

where

$T_v$ = the stroking time of the surge valve
$T_b$ = the stroking time of the block valve (from closed position to position that initiates surge)
$T_i$ = the integral (reset) time setting

## *8.2 Disturbance Time Interval*

The discussion up to this point has assumed that the disturbance time interval is longer than the control loop settling time. If the disturbance time interval is less than the loop period multiplied by 3, the disturbance will be either in or out of phase with existing oscillations. In-phase disturbances will accentuate and out-of-phase disturbances will diminish the existing oscillations. In either case, the oscillations will become noticeably distorted. The proportional band can be decreased and the derivative time can be increased to increase the speed of reaction and recovery before the next disturbance arrives. Also, the integral time can be increased to provide more averaging by the reset response of the disturbances. These guidelines agree with the stochastic control mode settings in Sood and Huddleston (1977). If the disturbance time interval becomes less than one fourth of the loop period, the disturbances are uncontrollable and can be considered as noise. The control mode settings must then be detuned per Section 8.3. Control loop noise can originate from uncontrollable disturbances, extraneous measurement effects, and electrical interference. A furnace pressure loop is an excellent example of significant noise resulting from uncontrollable disturbances (caused by fan pulsing, nonhomogeneous mixing, and interaction) and measurement noise (caused by velocity head and flow separation and vortices) (McMillan, 1980). The permissible amount of derivative time is drastically reduced and is essentially zero for loops with small time constants. Loops with large dead-time to time-constant ratios have derivative times much less than $0.125^*TD$ because the noise accentuated by derivative action is not sufficiently filtered by time constants within the loop. Even if the process has a large time constant, the rapid oscillation of the control valve due to noise may be undesirable from a maintenance viewpoint. The standard deviation of the control valve signal for proportional plus integral control can be estimated by Equation 8.4 from Fertik (1975).

$$E_v = E_n^* \left[\frac{100}{PB}\right]^* \left[\frac{TC_n}{TC_f + TC_n}\right]^{0.5} \tag{8.4}$$

$$TC_n = 0.8^* \left[\frac{T_z}{N_z}\right] \tag{8.5}$$

$$E_n = \frac{E_p}{8} \tag{8.6}$$

where

$E_v$ = standard deviation of the control valve signal
$E_n$ = standard deviation of the noise signal
$PB$ = proportional band of the controller
$TC_f$ = time constant of the measurement filter

$TC_n$ = dominant time constant of the exponentially correlated noise

$T_z$ = total time interval under study

$N_z$ = number of zero crossings of the noise signal in the time $T_z$

$E_p$ = peak-to-peak amplitude of the noise signal

About 100 zero crossings are needed ($N_z = 100$) to obtain a good estimate of the time constant of the noise signal. Equation 8.3 can be used to determine the size of the measurement filter time constant necessary to provide the desired degree of attenuation. To avoid excessive valve wear, the standard deviation of the valve signal should be less than the valve stroke resolution. If a measurement filter is not used, the noise time constant cancels out in Equation 8.4 and all of the desired attenuation must be achieved by increasing the proportional band. The increase in proportional band due to an increase in measurement time constant can be compared with the increase in proportional band necessary to attenuate unfiltered noise. The increase in integral time owing to the measurement filter must be included to estimate the increase in accumulated error. Noise attenuation by use of a measurement time constant instead of by use of just a large proportional band will typically yield a smaller peak error but a larger accumulated error. Measurement noise in a signal selector control system will cause an offset of the controlled variable from the set point. The offset can be reduced significantly by the use of integral-only control action in the override controllers. The offset increases with noise level (standard deviation) about three times as fast for PID override controllers than for PI override controllers (Weber & Zumwalt, 1979).

## 8.3
## *Furnace Pressure Loop Example*

*Given:*

(Information from Examples 4.6.4, 6.4.4, and 7.4)

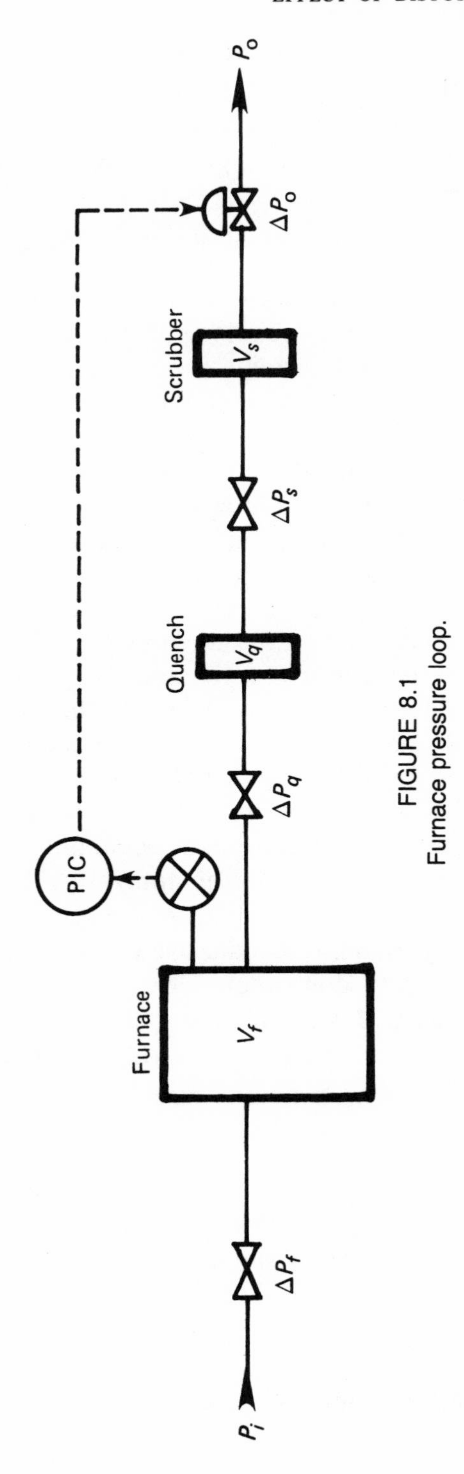


FIGURE 8.1
Furnace pressure loop.

(a) Set point is 5 inches w.c. gage.
(b) Measurement range is 0 to 10 inches w.c. gage.
(c) Furnace volume is 10,000 ft$^3$ ($V_f$).
(d) Quench volume is 1000 ft$^3$ ($V_q$).
(e) Scrubber volume is 1000 ft$^3$ ($V_s$).
(f) Furnace inlet flow resistance pressure drop is 2.5 inches w.c. ($\Delta P_f$).
(g) Quench inlet flow resistance pressure drop is 5 inches w.c. ($\Delta P_q$).
(h) Scrubber inlet flow resistance pressure drop is 10 inches w.c. ($\Delta P_s$).
(i) System outlet flow resistance pressure drop is 2.5 inches w.c. ($\Delta P_o$).
(j) Flue gas flow is 1000 scfm ($F_f$).
(k) Atmospheric pressure is 408 inches w.c. ($P_a$).
(l) Disturbance is a rapid 20 percent increase in inlet pressure ($\Delta L$).
(m) Inlet pressure (discharge of forced draft fan) is 15 inches w.c. ($P_i$).

(New information)

(n) Damper stroke resolution is 0.5 percent in control range.
(o) Total time interval under study is 5 minutes ($T_z$).
(p) Number of zero crossings is 100 ($N_z$).
(q) Peak-to-peak noise signal is 3.2 inches w.c. ($E_p$ = 32 percent).

*Find:* the new peak error for the *PB* necessary to attenuate the noise.

*Solution:*

(a) Calculate the new proportional band:

$$E_n = \frac{E_p}{8} = \frac{32}{8} = 4$$

$$TC_n = \frac{0.8 * T_z}{N_z} = \frac{0.8 * 300}{100} = 2.4 \text{ seconds}$$

$$E_v = E_n * \left[\frac{100}{PB}\right] * \left[\frac{TC_n}{TC_n + TC_m}\right]^{0.5} \tag{8.4}$$

$$PB = E_n * \left[\frac{100}{E_v}\right] * \left[\frac{TC_n}{TC_n + TC_m}\right]^{0.5}$$

$$PB = 4 * \left[\frac{100}{0.5}\right] * \left[\frac{2.4}{2.4 + 1.7}\right]^{0.5} = 612 \text{ percent}$$

(b) Calculate the new peak error:

$$E_x = \left[\frac{1.1 * PB}{100}\right] * E_o \tag{3.11}$$

$$E_x = \left[\frac{1.1 * 612}{100}\right] * 0.29$$

$$= 1.95 \text{ inches w.c.}$$

*Conclusions:* The noise increased the peak error by only 60 percent because the proportional band was already large because of the slow transmitter and valve. If the transmitter time constant had been set at its minimum ($TC_m = 0.2$), the peak error would have increased by an additional 25 percent but the accumulated error would have decreased by 5 percent due to a 30 percent decrease in loop period.

# CHAPTER IX
# Effect of Nonlinearities

## 9.1
## Variable Gain

Although the instrument engineer may choose a valve size and transmitter calibration such that the overall product of the loop steady-state gains is unity ($K_v*K_p*K_m = 1$), the individual gains frequently will be nonlinear and can be only approximated as linear for a small operating region. If the change in gain is small or the disturbance size is small, a simple adjustment of the proportional band based on the localized gains is sufficient. The localized steady-state gain is the slope of the output versus input curve at the operating point. If the curve changes with time (e.g., waste treatment pH titration curve), a family of curves is necessary for analysis. The valve gain for the installed characteristic of a equal percentage valve (see Figure 9.1) shows a relatively flat operating region near mid-scale. The width of this flat region increases as the percentage of system pressure drop allocated to the valve decreases (Buchwald, 1974).

Since many valves are sized to operate near midscale, the installed characteristic of the equal percentage valve in a liquid system provides an approximately linear valve gain even for moderately sized disturbances. The installed characteristic of a linear valve in a liquid system is quick opening and can only be assumed linear at midscale for extremely small disturbances. The installed characteristic of a conventional butterfly valve will show practically no gain above midscale (see Figure 9.2). The controller action will not have any effect on flow and the controller may wind up as a result of reset action. If the system frictional pressure drop changes with time due to fouling, a family of installed characteristic curves is needed.

Steady-state simulation can be used to compute the change in process output for a change in process input to estimate the

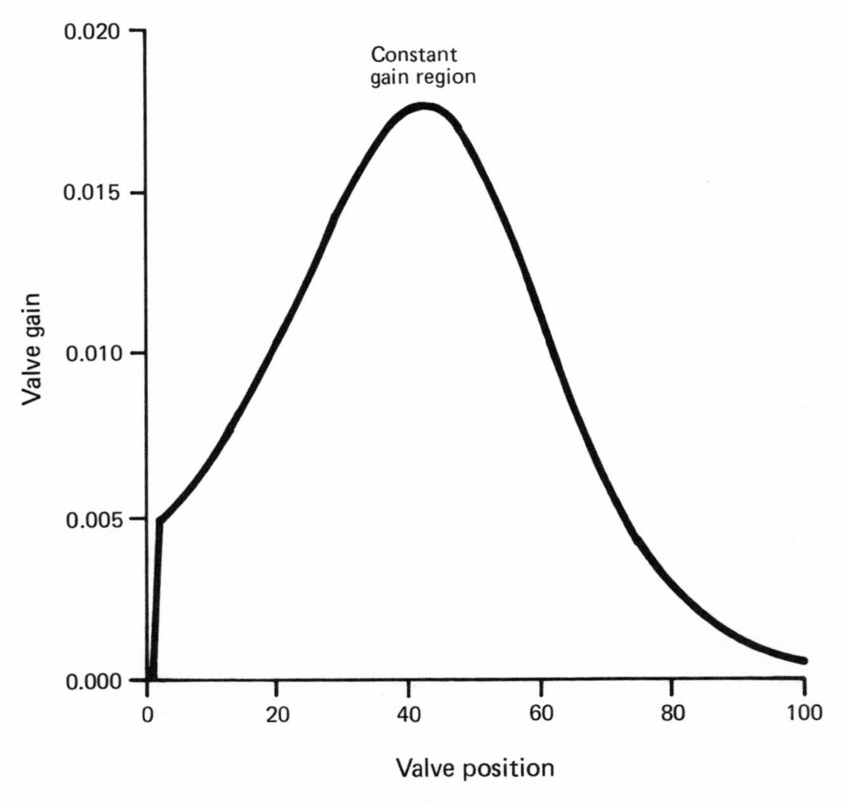


FIGURE 9.1
Installed valve gain of an equal percentage inherent characteristic.

process gain. The process gain can also be found by taking the derivative of the controlled process variable with respect to the manipulated variable (usually flow). Steady-state mass or energy balances provide an equation that can be used to estimate the steady-state process gain. The steady-state process gain, $K_p$, can be found in Appendix C for various processes (setting the time derivative to zero in the mass and energy balance equations and taking the derivative of the controlled variable with respect to the manipulated variable will yield $K_p$).

The process gain of a carbon monoxide control loop increases as the air flow (excess air) decreases. MeasureX® and Econics® carbon monoxide control systems linearize the process gain in their control algorithms. The process gain of an

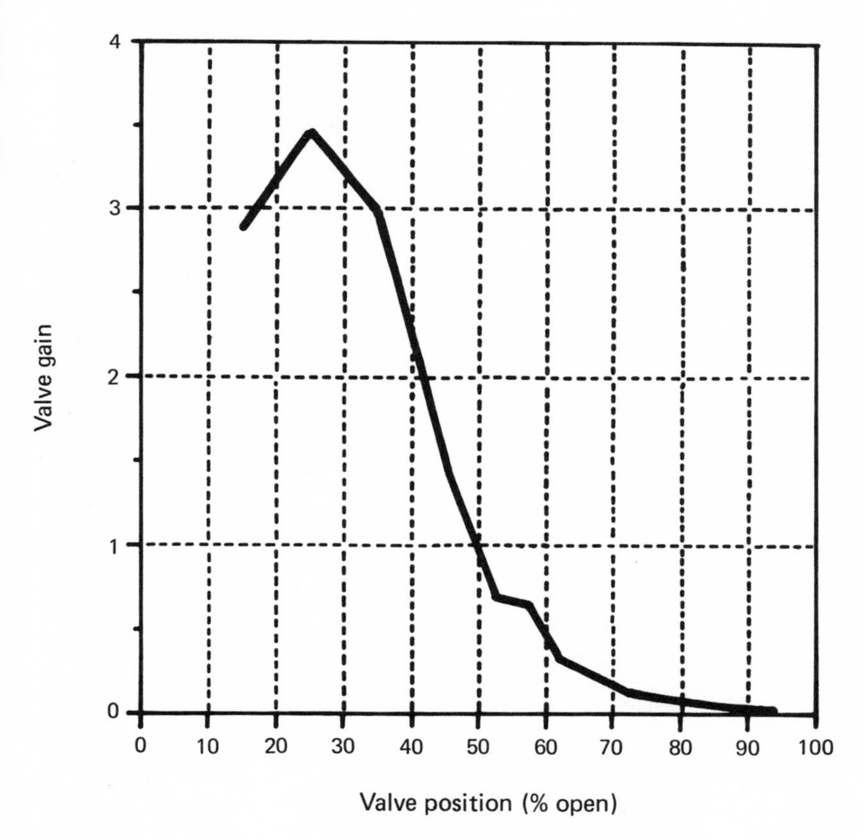


FIGURE 9.2
Installed valve gain of a conventional butterfly valve.

opacity control system has a minimum near the control point. The change in the minimum due to changes in reflectivity of the smoke particles makes it difficult for a control algorithm to correct for sign reversal unless the time constant and dead time of the response are predictable. The process gain of a heat exchanger decreases significantly as the flow increases, even though the heat transfer coefficient increases with flow. Figure 9.3 (a plot of Equation 9.1) shows that the negative process gain becomes more negative as either the hot fluid or the coolant flow decreases. An equal percentage characteristic will help compensate for the process gain nonlinearity (Shinskey, 1979). The use of a cascade loop from temperature to flow will accentuate the nonlinearity if a differential head flow transmitter is used (square root relationship causes measurement gain to decrease with flow).

$$\frac{\Delta T_{h2}}{\Delta W_c} = \frac{(T_{h1} - T_{c1})*\left[\frac{-2*W_h*C_h*C_c}{(2*W_c*C_c)^2}\right]}{D^2} \tag{9.1}$$

$$D = \left[\frac{W_h*C_h}{U*A}\right] + \left[\frac{1 + \frac{(W_h*C_h)}{(W_c*C_c)}}{2}\right] \tag{9.2}$$

where

$\frac{\Delta T_{h2}}{\Delta W_c}$ = process gain

$T_{h1}$ = hot fluid entrance temperature

$T_{h2}$ = hot fluid exit temperature

$W_h$ = hot fluid mass flow

$C_h$ = hot fluid heat capacity

$T_{c1}$ = coolant entrance temperature

$W_c$ = coolant mass flow

$C_c$ = coolant heat capacity

$U$ = overall heat transfer coefficient

$A$ = heat transfer area

Significant measurement nonlinearities occur for conductivity, ORP, pH, low-temperature zirconium oxide oxygen, and differential head flow transmitters. Conductivity is a complex function of concentration and can have localized maxima and minima with the associated gain reversals due to changes in the activity coefficient and to ion pair formation (see Figure 9.4) (McMillan, 1978). The ORP measurement gain depends on both concentration and pH (McMillan, 1980). The pH measurement gain of a strong acid and base neutralization (acid and base are completely ionized) changes by a factor of 10 for each pH unit deviation from neutrality. The pH measurement gain at 7 pH is about seven orders of magnitude larger than the pH measurement gain at 0 pH (see Equation 9.3) (Shinskey, 1973). The pH measurement gain for multiple reagents with different ionization constants will depend not only on pH, but also on the

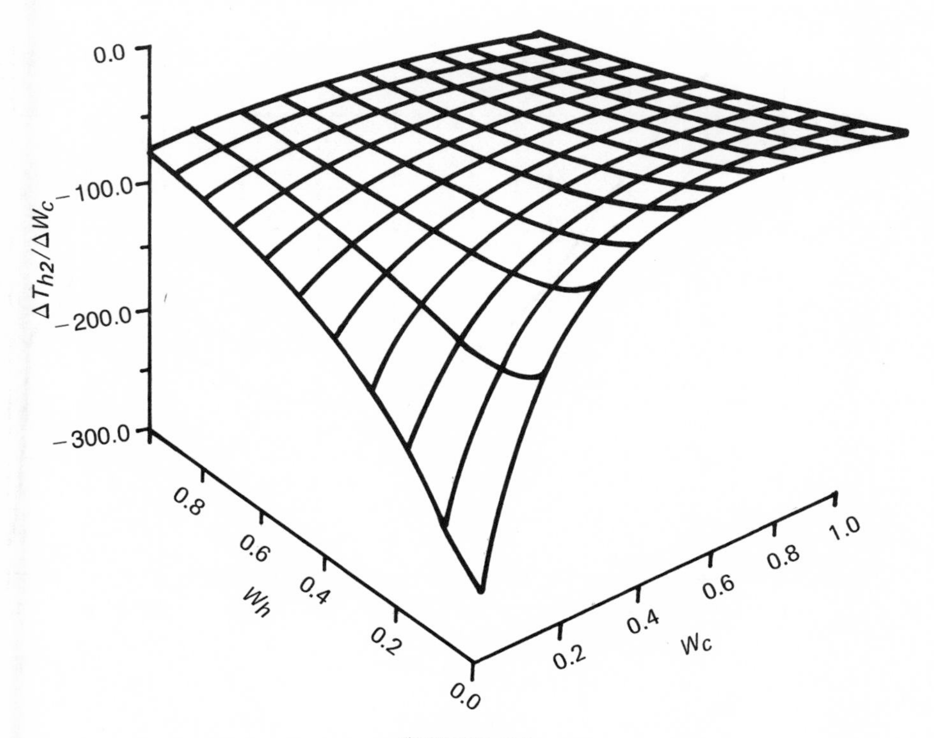


FIGURE 9.3
Process gain for a heat exchanger.

relative flow of each reagent (McMillan, 1979). For a strong acid and strong base (no buffering):

$$\frac{\Delta pH}{\Delta X} = \frac{-0.434}{10^{(-pH)} + 10^{(pH - 14)}} \tag{9.3}$$

where

$\frac{\Delta pH}{\Delta X}$ = measurement gain (pH per normality of excess ions)

$pH$ = pH of the solution

If the change in control valve, process, or measurement gain is large in the operating region, the dead time or time constant

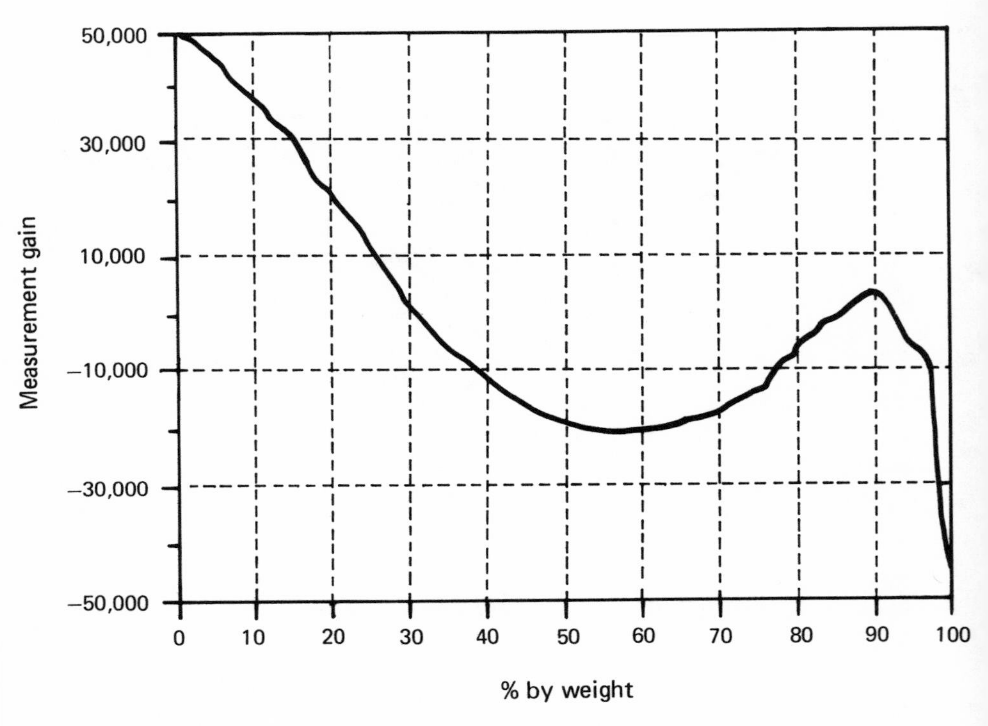


FIGURE 9.4
Conductivity measurement gain for sulfuric acid concentration.

should be corrected. If the steady-state gain increases with the error, the effective dead time will be larger than the actual dead time. If the steady-state gain decreases with error (e.g., pH control at neutrality), the effective time constant is smaller than the actual time constant. In either case, the proportional band should be increased accordingly. If the dead time and time constant are measured graphically from a field recording, the measured dead time and time constant will be the effective dead time and time constant.

## *9.2*
## *Variable Time Constant and Dead Time*

Time constants and dead times change when the gain nonlinearity changes; when the energy balance's mass flows, heat capacities, or heat transfer coefficients change; and when the mass balance's hold-up volumes or flows change. The change in the mass balance time constants and dead times is easy to predict and to compensate for in a microprocessor-based controller. The hold-up volume, calculated from level or pressure measurements and equipment dimensions, and the measured flows can be substituted into the equations in Appendix E for liquid composition control and in Appendix G for gas pressure control. The change in the ultimate period and the associated controller mode settings can be calculated from the equations in Chapter 4. If the agitation flow is much greater than the throughput flow, the integral and derivative times of the controller need not be corrected for throughput flow changes in self-regulating liquid composition control since the change in dead time is negligible per Equation E1 (the ultimate period is equal to about four dead times for small dead times and large time constants per Equation 4.5).

# 9.3 Inverse Response

If the initial response of the process is in a direction opposite to the final response, the process has inverse response. Inverse response does not have to result from a process nonlinearity. It can also be the result of competing effects of two processes or control systems with different dynamics. Inverse response occurs in boiler drum and distillation column sump level control. Inverse response can also originate from too large a feedforward gain, too high a lead time setting, or too low a lag setting, or from interaction with other loops. The current practice is to approximate the duration of the opposite response as additional dead time. However this ignores the controller response to the time duration and the amplitude of the peak in the wrong direction. The effect of reset action is small if the time duration of this peak is short compared with that of succeeding peaks. The effect of rate action is small if the derivative setting was reduced to avoid reaction to the more rapidly changing initial peak. The effect of controller reaction to the first peak can be

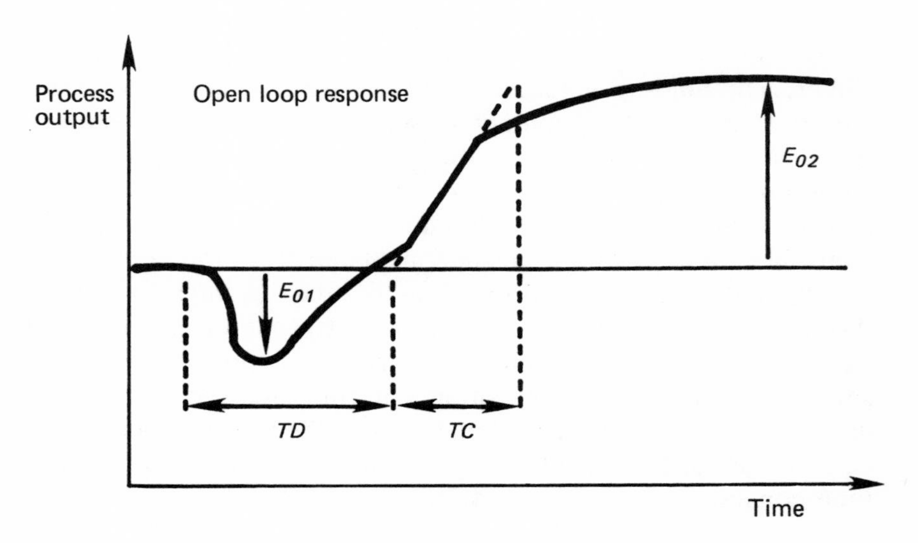


FIGURE 9.5
Inverse response of a self-regulating process.

included by the addition of the controller's response to the first and second peaks, which yields Equations 9.4–9.7 for self-regulating and integrating processes (the time duration of the initial peak should be added in as dead time for the calculation of the ultimate period—see Figure 9.5).

For self-regulating processes:

$$E_x = \left[\frac{1.1*K_g*T_u}{2\pi*TC}\right]*(E_{o1} + E_{o2}) \tag{9.4}$$

$$E_i = \left[\frac{K_g*T_u}{2\pi*TC}\right]*0.5*T_u*(E_{o1} + E_{o2}) \tag{9.5}$$

For integrating processes with $T_u/TC < 4$:

$$E_x = \left[\frac{1.1*K_g*T_u}{2\pi*TC}\right]*\left[E_{o1} + \frac{T_u}{2\pi}*E_{o2}\right] \tag{9.6}$$

$$E_i = \left[\frac{K_g*T_u}{2\pi*TC}\right]*0.5*T_u*\left[E_{o1} + \frac{T_u}{2\pi}*E_{o2}\right] \tag{9.7}$$

where

$E_x$ = peak (maximum) error
$E_i$ = accumulated (integrated) error
$K_g$ = controller gain factor ($K_g = 1.5$)
$T_u$ = ultimate period (includes the inverse response dead time)
$TC$ = dominant time constant of the loop
$E_{o1}$ = open loop error of the first peak
$E_{o2}$ = open loop error of the second peak

## 9.4 *Waste Treatment pH Loop Example*

*Given:*

(Information from Examples 4.6.1 and 6.4.1)

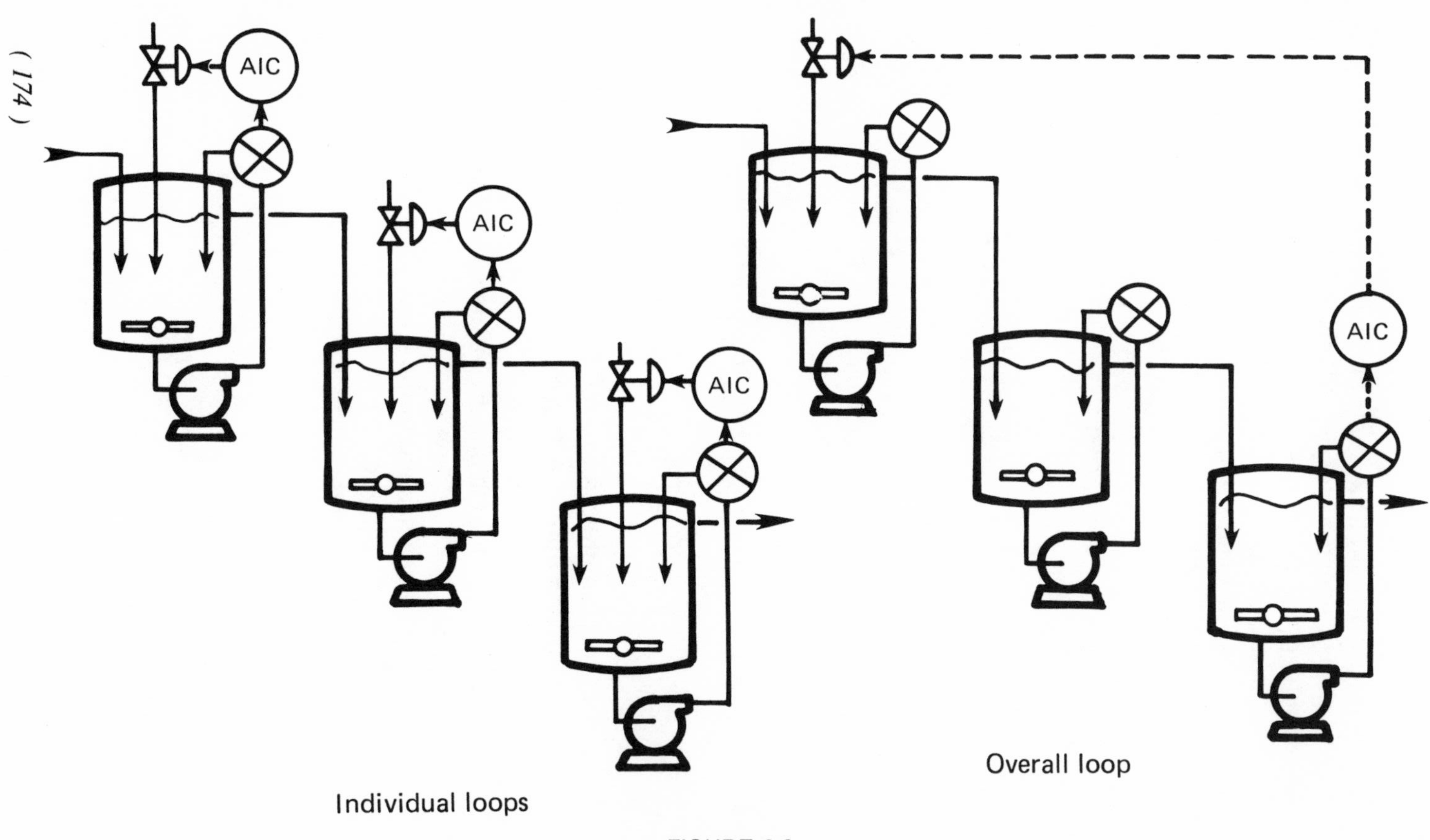


FIGURE 9.6
Waste treatment pH loop.

(a) Set point is 7 pH.
(b) Measurement range is 0 to 14 pH.
(c) Minimum influent flow is 10 gpm ($F_{il}$).
(d) Normal influent flow is 22 gpm ($F_{in}$).
(e) Maximum influent flow is 100 gpm ($F_{ih}$).
(f) Influent concentration is 32 percent by weight HCl—10.17 normality ($C_i$).
(g) Influent disturbance is rapid 20-gpm increase in flow ($\Delta L$).
(h) Reagent concentration is 20 percent by weight NaOH—7.93 normality ($C_r$).
(i) Vertical tank liquid volume is 1000 gallons ($V$).
(j) Axial blade agitator diameter is 1 foot ($D$).
(k) Axial blade agitator speed is 120 rpm ($N_s$).
(l) Axial blade agitator discharge coefficient is 1 ($N_q$).

(New information)

(m) Nonlinear controller with notch gain breakpoints set at 1 and 13 pH.

*Find:* the increase in peak error for an individual tank.
*Solution:*
(a) Calculate the effective time constant seen by the controller:

$$pH_f = -\log(2) = -0.3$$

$$pH_e = 0.63^*(pH_f - pH_i) + pH_i$$

$$= 0.63^*(-0.30 - 7) + 7 = 2.4$$

$$C_e = 10^{(-pH_e)} = 0.004$$

$$C_e = \left[1 - e^{-TC_e/TC}\right]^*(C_f - C_i) + C_i$$

$$\frac{TC_e}{TC} = -\ln\left[1 - \frac{C_e - C_i}{C_f - C_i}\right]$$

$$\frac{TC_e}{TC} = -\ln\left[1 - \frac{0.004 - 0}{2 - 0}\right] = 0.002$$

$$TC_e = 0.002^*TC = 0.002^*19 = 0.04 \text{ minute}$$

(b) Calculate the notch gain of a nonlinear controller:

$$\frac{\Delta pH}{\Delta C} = \frac{-7.3}{2} = -3.65 \text{ (gain if measurement was linear)}$$

$$K_m = \frac{-0.434}{10^{(-pH)} + 10^{(pH - 14)}} \qquad (9.1)$$

$$K_m = \frac{-0.434}{10^{-2}} = -4.3 \text{ at 1 pH}$$

$$K_m = \frac{-0.434}{10^{-7} + 10^{(7 - 14)}}$$

$$= -2{,}170{,}000 \text{ at 7 pH}$$

$$K_{db} = \frac{-4.3}{-2{,}170{,}000}$$

$$= 0.000002 \text{ notch gain for stability at 7 pH}$$

[Calculated notch gain is 10,000 times smaller than that typically available in a nonlinear controller ($K_{db} > 0.02$). Therefore, the loop will oscillate between approximately 3 and 11 pH continuously.]

(c) Calculate the new ultimate period:

$$T_u = 2*\left[1 + \left[\frac{TC}{TC + TD}\right]^{0.65}\right]*TD$$

$$T_u = 2*\left[1 + \left[\frac{0.04}{0.04 + 3.1}\right]^{0.65}\right]*3.1 \qquad (4.5)$$

$$= 6.6 \text{ minutes}$$

(d) Calculate the new peak error:

$$PB = K_g*100*K_v*K_p*K_m \qquad \text{since } TC \ll T_u \qquad (4.12)$$

$$PB = 1.5*100*1*1*(-4.3)$$

$$= -645 \text{ percent (minus sign denotes reverse action)}$$

$$E_x = \left[\frac{1.1*PB}{1.00*K_v*K_p*K_m + PB}\right]*E_o \qquad \text{since } TC \ll T_u$$

$$E_x = \left[\frac{1.1*645}{430 + 645}\right]*2 = 1.45 \text{ normality} \tag{3.10}$$

$$= 1.45 \text{ normality}$$

$$E_x = 7 - (-\log(1.45)) = 7 + 0.16 = 7.16 \text{ pH}$$

*Conclusions:* Immeasurably small disturbances will cause the loop for the first tank to oscillate continuously between 3 and 11 pH. The peak error for the given disturbance is 7.16 pH for the first tank. If a 0 to 14 pH scale were chosen, the measurement would go offscale (−0.16 pH). Measurement signal linearization not only would eliminate the notch-gain-induced oscillations, but would also restore the effective time constant to its original value of 19 minutes (McMillan, 1981a).

CHAPTER X

# *Effect of Interaction*

If control actions from each loop affect the others, the loops are said to interact. A steady-state gain matrix can be constructed if the controllers are placed in manual, the manual output of each is individually changed, and the final value of every controller measurement is recorded for each change. If the controller measurements are first passed through the inverse of this gain matrix en route to the controllers, the loops will be fully decoupled per steady-state data. If the gain matrix is multiplied by the corresponding elements of its inverse matrix, the result is the relative gain matrix. The best pairing of measurements and control valves are those combinations whose relative gain elements are closest to unity. It is desirable that all other possible combinations have relative gain elements close to zero. A relative gain element between zero and one means that the net effect of the other loops is to change the measurement signal in the same direction as the control valve signal (cooperation). A relative gain element greater than one means that the net effect of the other loops is to change the measurement signal in the opposite direction to the control valve signal (conflict). A relative gain element less than zero means that the net effect is also conflict but the interaction dominates and causes the gain to change sign. Thus, if this loop is stable when the other loops are on manual, it probably is not stable when the other loops are on automatic, and vice versa. Even if the loop is stable, the measurement will have to overshoot the set point to reach steady state.

The relative gain matrix provides a useful tool for estimating the steady-state interaction between control loops. However the actual interaction depends upon the relative dynamics of the interacting loops. A relative gain analysis for the furnace pressure control loop shows that the net effect of the other loops is dominant opposition as a result of the severe interaction of the steady-state material and energy balances. The furnace pres-

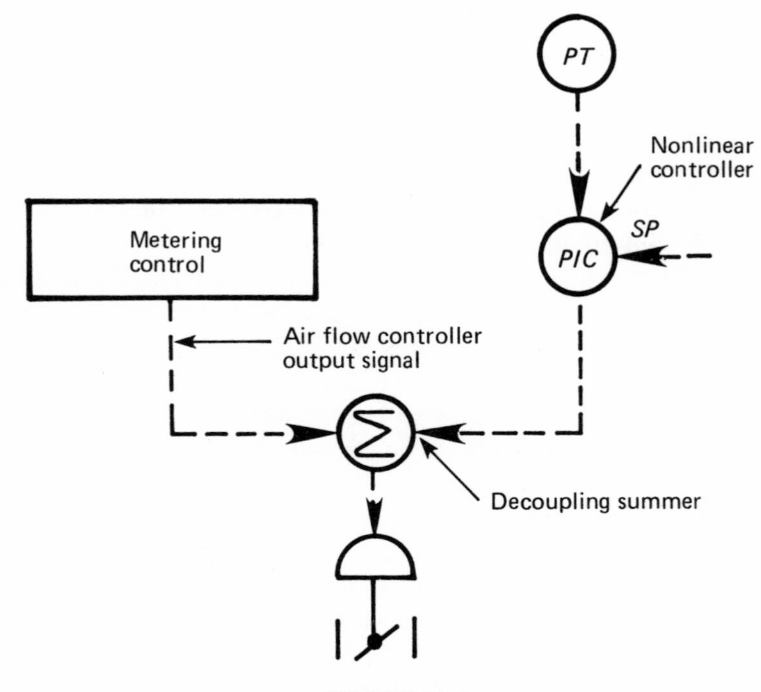


FIGURE 10.1
Half-decoupling of furnace pressure control.

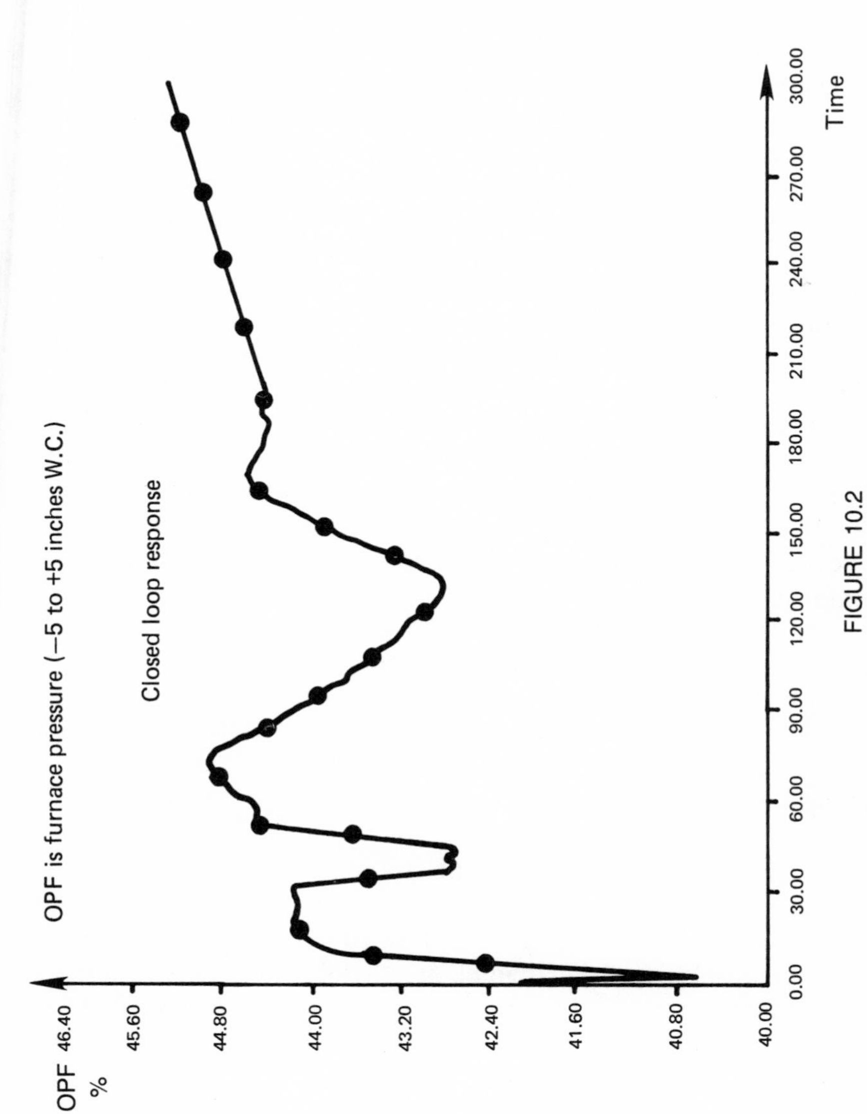


FIGURE 10.2
Pulverized coal-fired boiler furnace pressure response.

sure is controllable only because the dynamics of the gas loops are faster than the steam loops (McMillan, 1980). The furnace pressure control loop is typically decoupled from the faster combustion air flow control loop. Figure 10.1 shows that the decoupling may simply require a summer on the signal to the control valve of the slower loop to add the controller output from the faster loop (combustion air) to the slower loop (furnace pressure).

If full decoupling is required but not implemented, the proportional band of both loops will have to be increased to reduce reaction to the interaction. For extreme cases of interaction, increasing the proportional band is not sufficient and one of the controllers must be placed in manual. If half decoupling is required but not implemented, the proportional band of the slower loop will have to be increased to reduce reaction to disturbances from the faster loop. Increasing the proportional band unfortunately increases the error due to disturbances originating within the loop. If a loop is noisy and has to have its proportional band increased significantly, decoupling may not be necessary. Severe interaction may manifest itself as inverse response and high-frequency uncontrollable disturbances. Figure 10.2 shows noise in the closed loop response of furnace pressure control as a result of interaction (measurement noise due to turbulence of the combustion gas or atmospheric drafts was not included).

The tuning of interacting loops can be extremely tedious because the required tuning parameters change as interacting loops are commissioned. The faster loops or the loops least affected by the other loops should be tuned first. After each loop is tuned, it must be left in automatic so that its effect on subsequent loops is included. The slower loops or the loops with the most interaction should be tuned last. If stability becomes a problem, the sequence of controllers tuned to that point should be repeated, but with the previously tuned controllers left in automatic. The sequence may have to be repeated successively until the controllers are detuned enough for stability. A good example of the application of this technique has been documented for a once-through supercritical pressure generating unit (Morse, 1977).

CHAPTER XI

# *Effect of Advanced Control Algorithms*

## *11.1 Cascade Control*

In cascade control, there are several conventional feedback controllers and feedback measurements but only one manipulated variable. The output of each controller is the set point of another controller, except for the innermost controller. Figure 11.1 shows the most common type of cascade; it consists of two feedback controllers, two feedback measurements, and one control valve. The inner controller is also known as the "secondary" or "slave" controller and the outer controller is also known as the "primary" or "master" controller. If the control valve had a positioner, a triple cascade would actually exist (the positioner is a high-gain proportional-only controller).

Cascade control improves loop performance by rapidly correcting for the disturbance in the inner loop and using the outer loop time constant to filter the resulting inner loop oscillation. The improvement in performance is greatest when the inner loop controlled variable is selected so that the dead time to

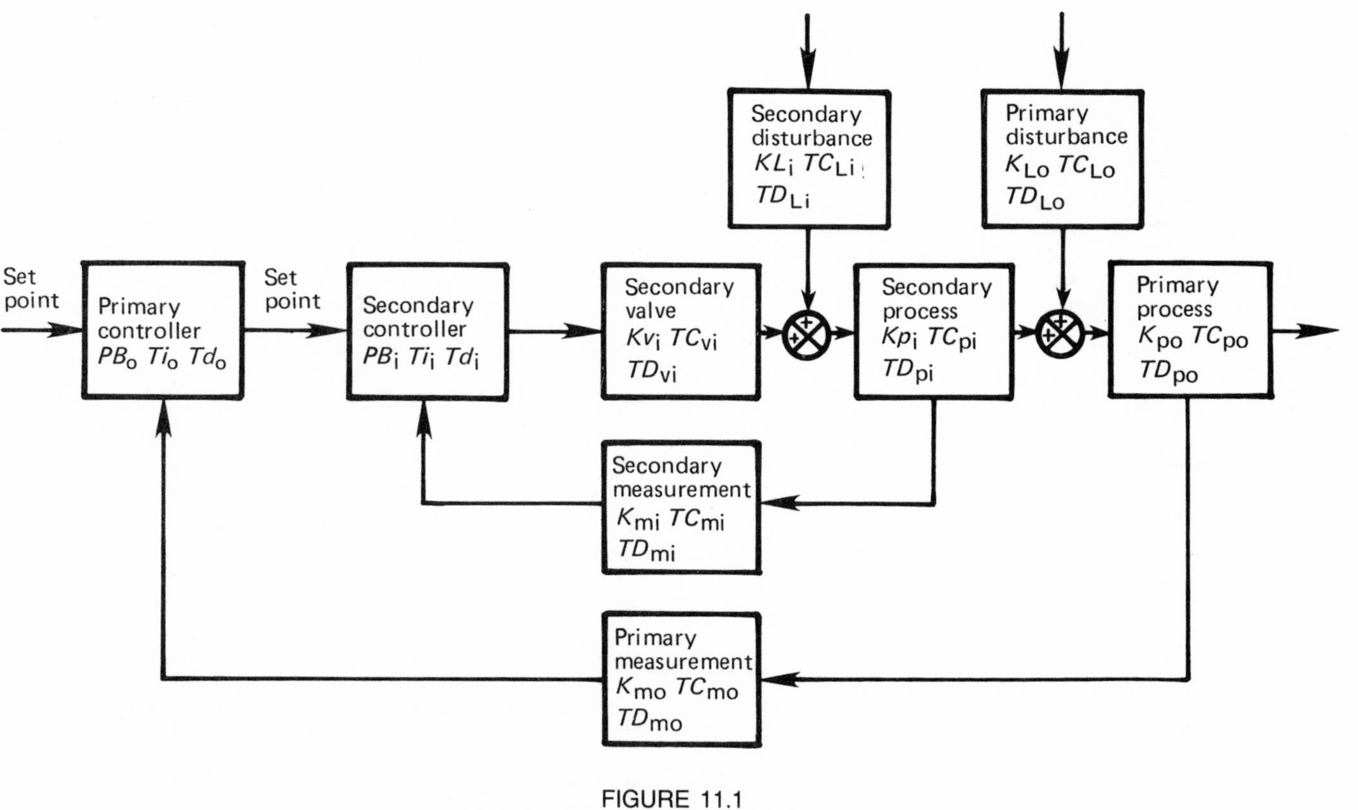


FIGURE 11.1
Block diagram of cascade control loop.

source of the disturbance is small (the ultimate period of the inner loop is then small), the inner loop process time constant is large for a self-regulating process (the proportional band of the inner controller is then small), and the time constant of the outer loop is large (the filtering effect of the outer loop is then large). These effects are illustrated by Equations 11.1 and 11.2, which approximate the quantitative reduction in peak error for inner loop disturbances by the use of cascade control (McMillan, 1982).

For a self-regulating inner and outer process:

$$E_{xr} = \left[\frac{PB_i}{PB_s}\right] * F_o * \left[\frac{K_{po} * K_{mo}}{K_{mi}}\right] \tag{11.1}$$

For a non-self-regulating inner or outer process:

$$E_{kr} = \left[\frac{PB_i}{PB_s}\right] * F_o * \left[\frac{PB_o}{100 * K_{mi}}\right] \tag{11.2}$$

For $TU_i < TC_i$:

$$F_o = \frac{TU_i}{\pi * TC_i} \tag{11.3}$$

where

$E_{xr}$ = ratio of outer to single loop peak error ($E_{xo}/E_{xs}$)
$PB_o$ = proportional band of the cascade outer loop
$PB_i$ = proportional band of the cascade inner loop
$PB_s$ = proportional band of the original single loop ($PB_o < PB_s$ due to reduction in loop period by cascade)
$F_o$ = filter factor of the outer loop
$TU_i$ = ultimate period of the inner loop
$TC_o$ = time constant of the outer loop
$K_{po}$ = steady-state process gain of outer loop
$K_{mo}$ = steady-state measurement gain of outer loop
$K_{mi}$ = steady-state measurement gain of inner loop

The above equations were developed from the application of Equations 3.10 and 6.2, except that the filtering effect was halved. Dynamic simulation results were used to verify the accuracy of the equations. (See Appendix G for a listing of the ACSL dynamic simulation program.)

The appearance of the gain terms in equations 11.1 and 11.2 is misleading. These gains have no net effect on the peak error ratio because a change in one of these gains will result in a proportional change in the appropriate proportional band so that the effect cancels out (see Equations 4.11, 4.16, and 4.22).

Equation 11.1 for a self-regulating inner and outer loop does not include the outer loop proportional band. Nearly all of the decrease in peak error for inner loop disturbances is provided by the proportional band of the inner loop and the filtering by the time constant of the outer loop. If the inner loop period is less than the outer loop period, approximately the same peak error is observed whether the outer controller is on automatic or not. The accumulated error is, of course, much larger; also, there is no guarantee that the outer measurement will return to set point when the outer controller is on manual. Dynamic simulation results plotted in Figure 11.2 show that the peak error ratio increases exponentially with a decrease in the inner- to outer-time-constant ratio. The inner loop proportional band increases, the single loop proportional band decreases, and the inner loop ultimate period decreases for a decrease in the inner loop time constant. The decrease in ultimate period of the inner loop is small so that the decrease in the filter factor is small. The increase in the ratio of the inner to the single loop proportional band is much larger.

Equation 11.2 for a non-self-regulating process in the inner or outer loop contains the outer loop proportional band. When the inner loop is non-self-regulating, the inner loop ultimate period is much larger than the ultimate period of the outer loop, except for an extremely small inner loop dead time and time constant. Consequently the outer loop controller has time to react to reduce the offset. When the outer loop is non-self-regulating, any offset caused by the inner loop oscillation grows until corrected by the outer loop controller. Thus, in both cases, the proportional band of the outer loop has an effect on the peak error ratio.

Since there are two types of non-self-regulating processes (integrating and runaway), and the occurrence of a non-self-regulating process in both the inner and outer loops is rare, there are four combinations of interest. Dynamic simulation results are plotted in Figures 11.3–11.6 for the four combinations. Figures 11.2–11.6 all used the same base values of dead time and time constant. The outer loop dead time and time constant were fixed whereas the inner loop dead time and time

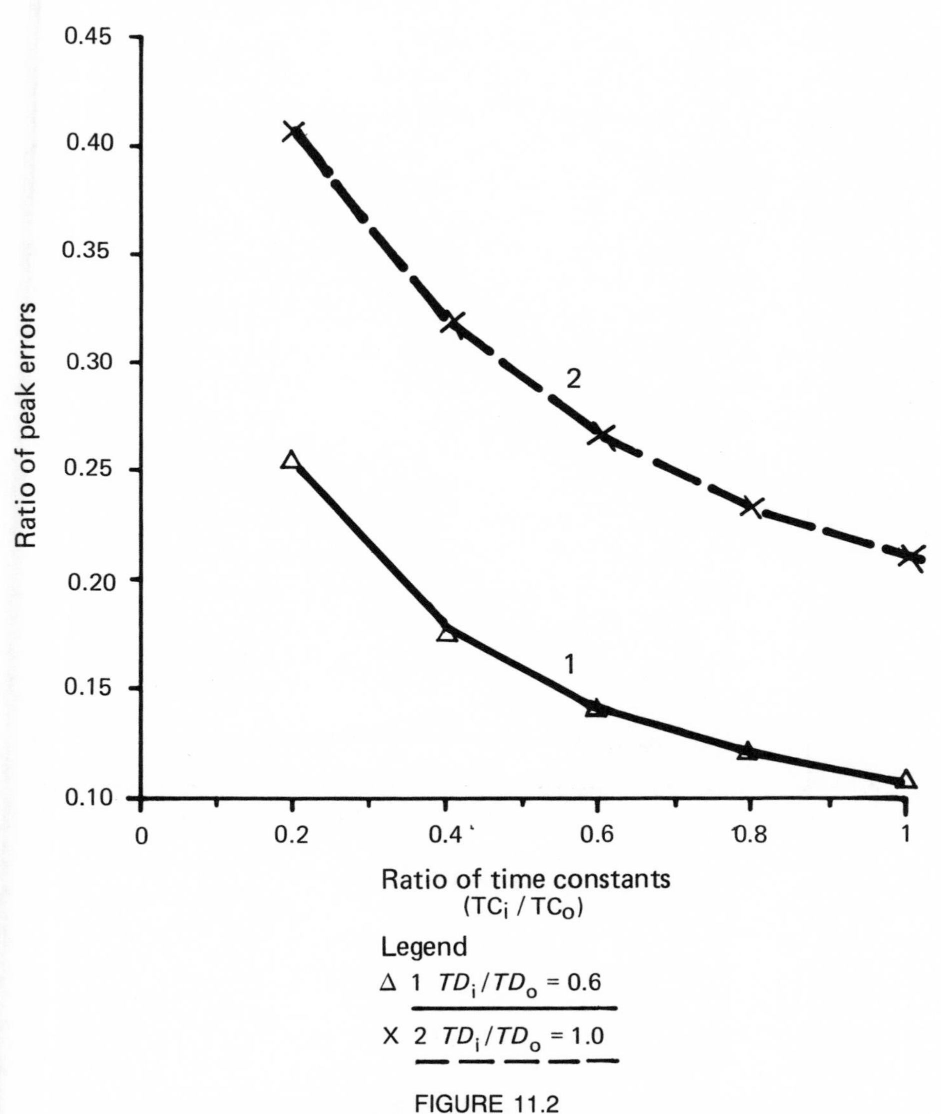


FIGURE 11.2
Peak error ratio for self-regulating inner and outer processes.

constant were varied. The PID controllers were tuned per the equations in Chapter 4. The effect of the inner loop response on the tuning of the outer controller was included by adding 90 percent of one fourth of the inner loop period (23 percent) as dead time to the outer loop dynamics. The simulation results show that about 72 percent of one fourth of the inner loop

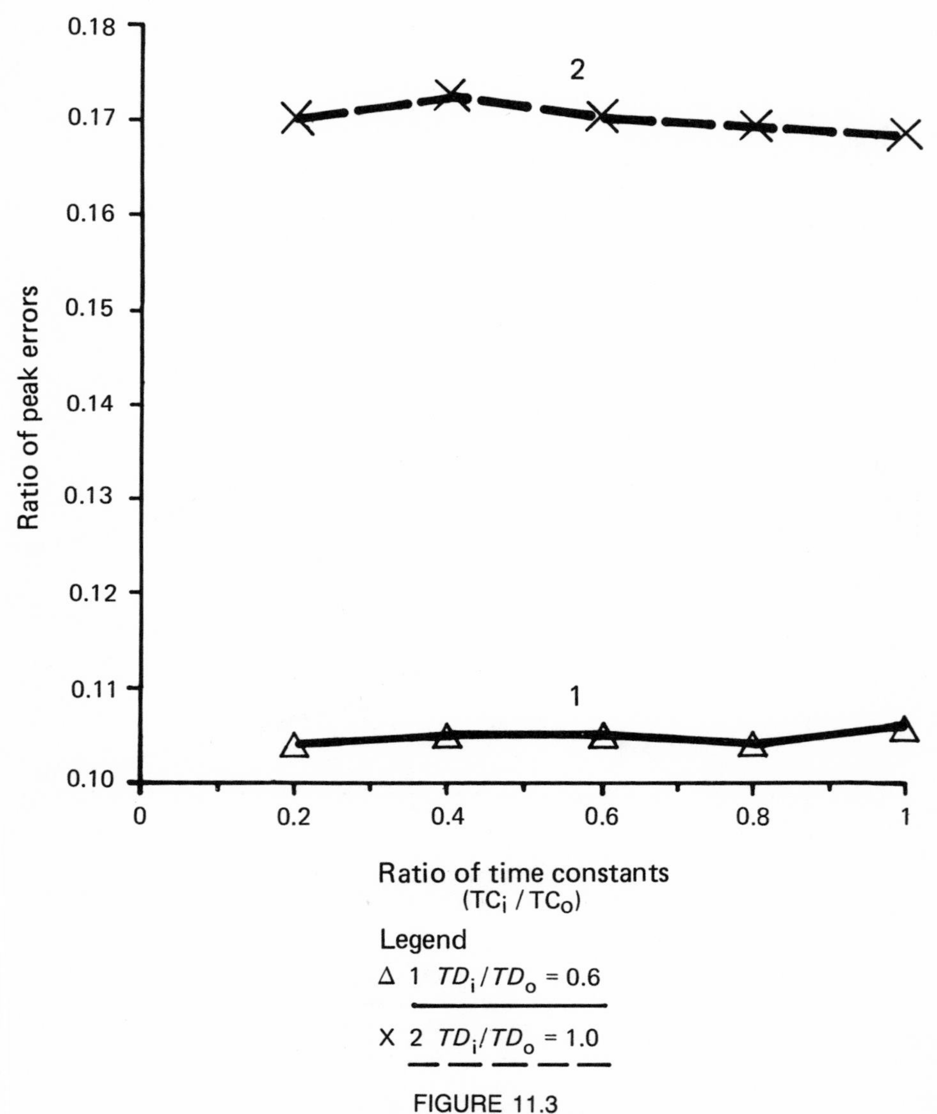


FIGURE 11.3
Peak error ratio for integrating inner processes.

period (18 percent) would have been more accurate when the outer loop had a self-regulating process.

Figure 11.3 for an inner integrating process shows that the peak error ratio is smaller than that in Figure 11.2 but there is negligible change with the inner- to outer-loop time-constant ratio. A decrease in the inner loop time constant decreases the

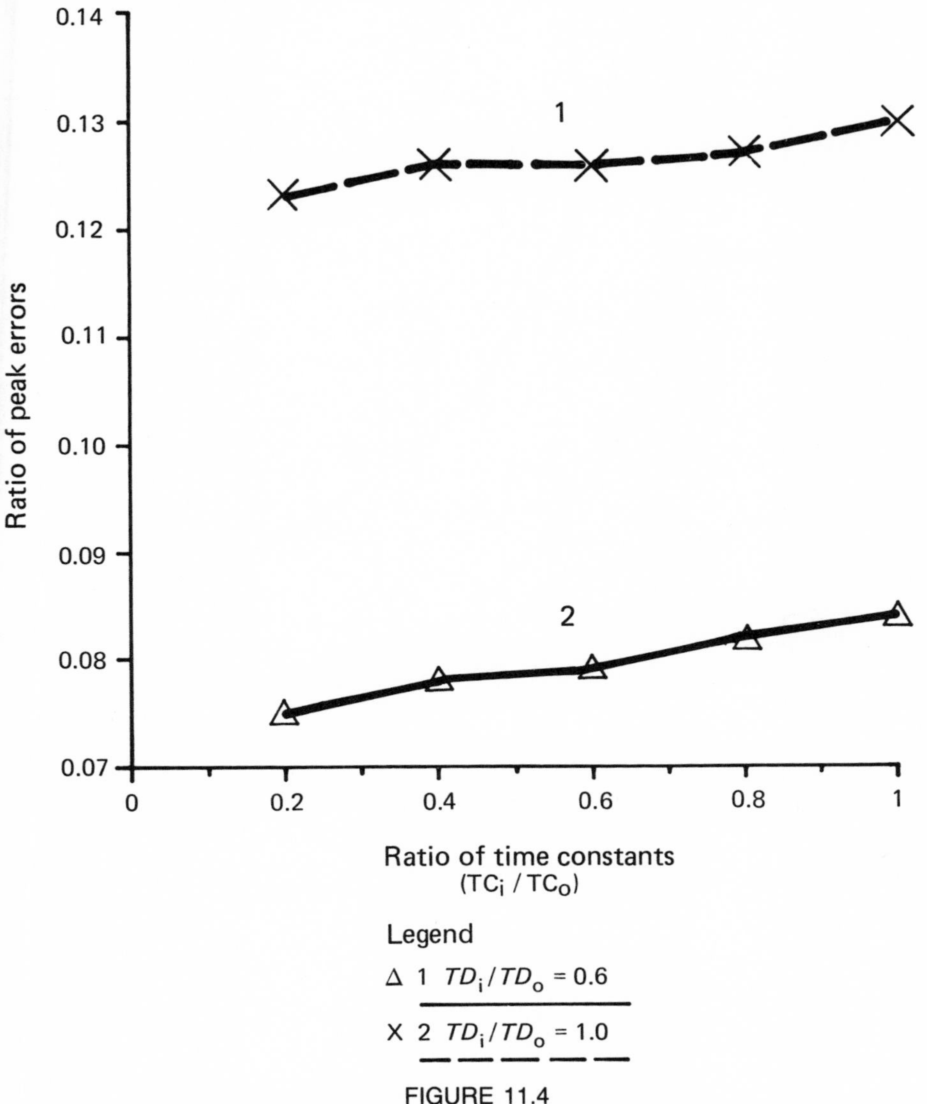


FIGURE 11.4
Peak error ratio for runaway inner processes.

proportional band of a loop with a non-self-regulating process (the opposite of the effect for a loop with a self-regulating process). Thus the changes in the proportional bands of the inner and outer loop tend to cancel out. The filter factor is approximately equal to one because the inner loop period is so large for non-self-regulating processes. The remaining term, the pro-

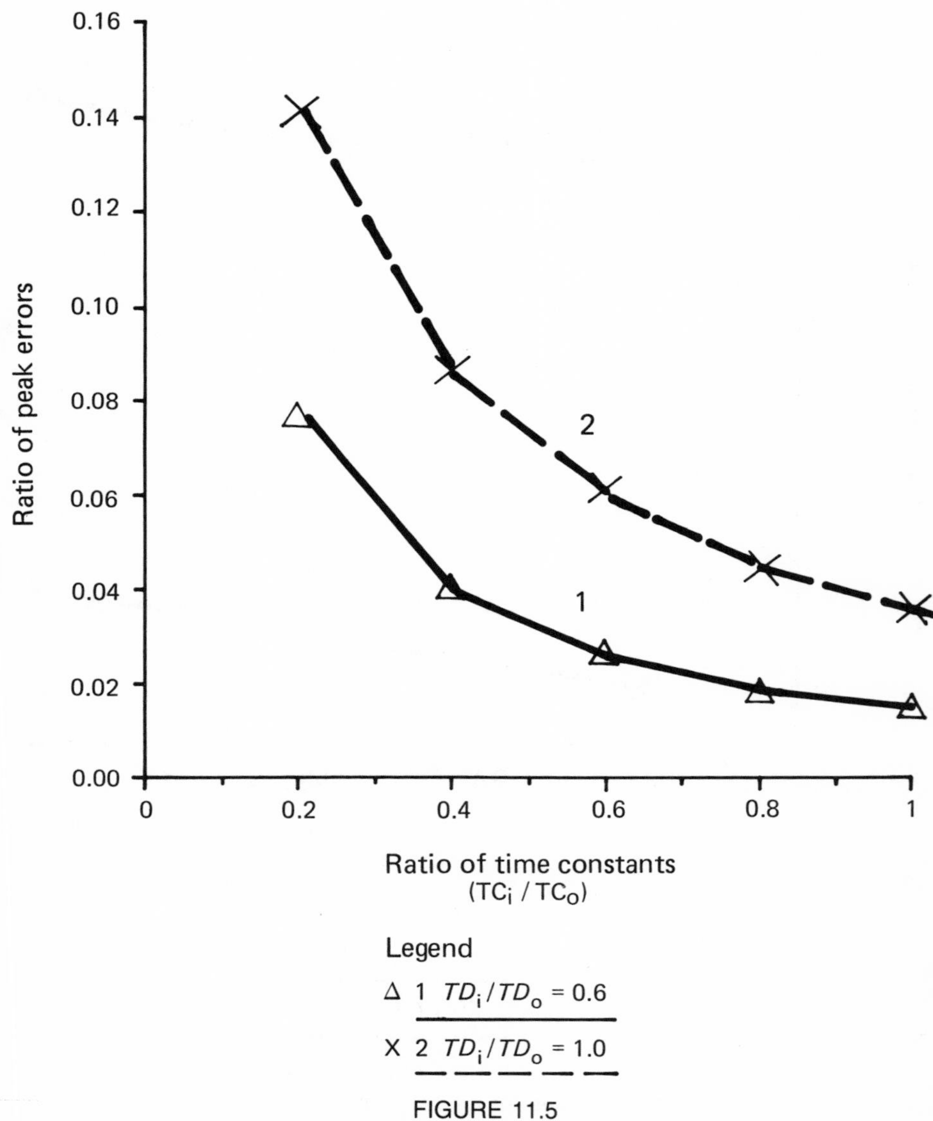


FIGURE 11.5
Peak error ratio for integrating outer processes.

portional band of the outer loop, is constant since the time constant and dead time of the outer loop were constant. Figure 11.4 for a inner runaway process shows slightly smaller peak errors and a slight decrease in the peak error ratio for a decrease in the inner- to outer-time-constant ratio. The decrease in the proportional band of the inner loop is slightly greater than the

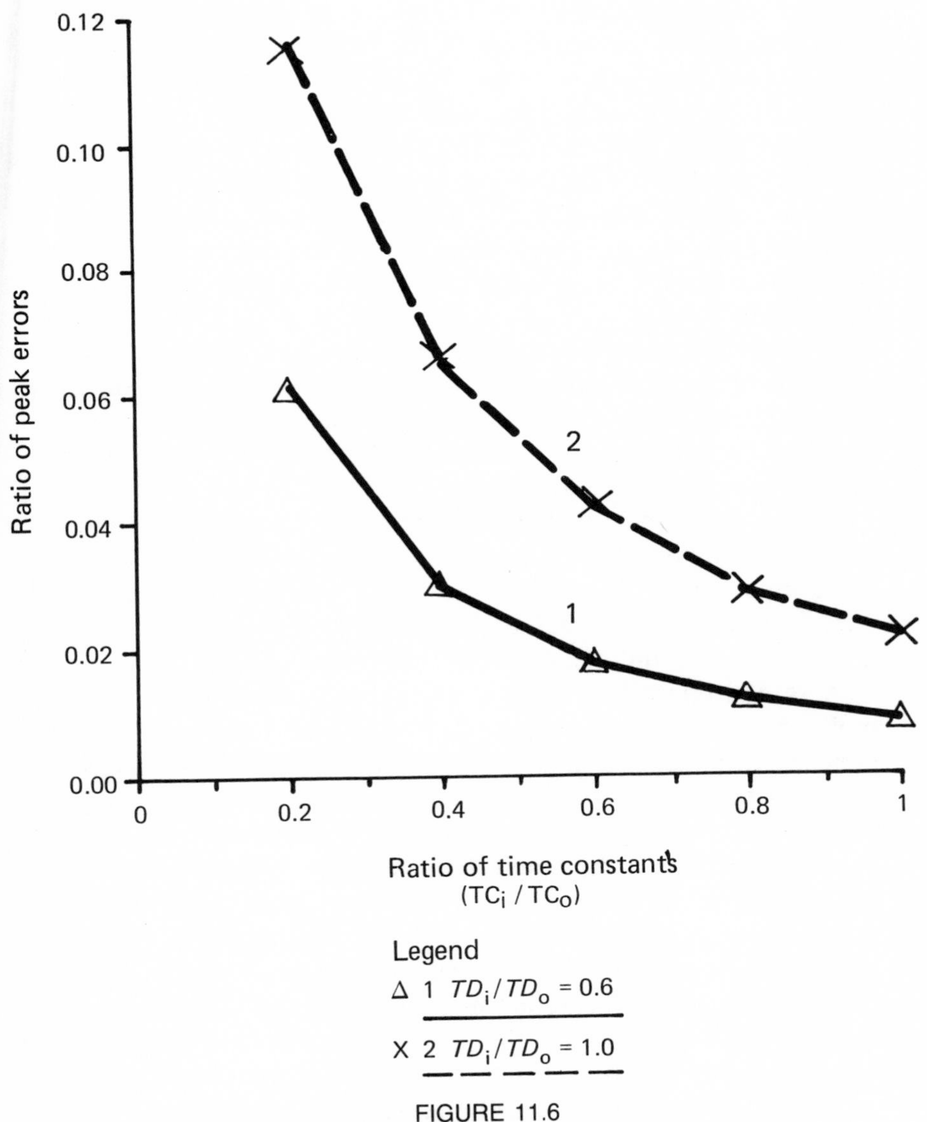


FIGURE 11.6
Peak error ratio for runaway outer processes.

increase in the proportional band of the single loop for a decrease in the time constant ratio. Figures 11.5 and 11.6, for an outer integrating and runaway process, respectively, show the same trend as Figure 11.2, which also had an inner self-regulating process except that the peak error ratios are much smaller and the exponential rise for small time constant ratios is faster.

The accumulated error for the outer loop becomes essentially zero because of the tendency of the outer loop oscillations to take longer to dissipate even though the second peak is one fourth of the first peak and larger on the opposite side of the set point to the disturbance. The positive error of the first half cycle is canceled out by the negative error of subsequent cycles. Thus the use of Equation 3.4 greatly overestimates the accumulated error for the outer loop. This great decrease in accumulated error may be an important criterion for the justification of cascade control.

The enclosure of some of the time constants and dead times of the process in a inner loop reduces the ultimate period for the outer controller so that its proportional band and integral time can be decreased. This additional effect will reduce errors that originate in the outer loop.

A process with an isolatable integrating or runaway response should be enclosed in a inner loop whenever possible (e.g., a level inner loop for reactor residence time control) because enclosing a non-self-regulating process in a inner loop makes it appear to the outer loop to be self-regulating.

Cascade control reduces the peak and accumulated errors more than dead-time compensators for inner loop load disturbances, if the dead time is the result of time constants in series. For self-regulating processes in both the inner and outer loops, the second largest time constant, which as equivalent dead time increased errors in the single loop, is used to decrease errors as the largest time constant in the inner loop. For non-self-regulating processes in the inner loop, the largest or second largest time constant, which as equivalent dead time increased errors in the non-self-regulating process of the single loop, is used to decrease errors as the largest time constant in the self-regulating process of the outer loop. For non-self-regulating processes in the outer loop, the largest or second largest time constant, which as equivalent dead time increased errors in the non-self-regulating process of the single loop, is used to decrease errors as the largest time constant in the self-regulating process of the inner loop. Thus a term detrimental to loop performance that creates dead time not only is eliminated, but is converted to a beneficial term by the proper use of cascade control. All secondary time constants in a self-regulating process and all time constants (negative feedback) in a non-self-regulating process have a detrimental effect due to a degree of created dead time. However the largest time constant in a self-regulating process has a beneficial effect due to a degree of filtering.

To maximize the improvement in loop performance, the inner loop measurement should be selected so that:

1. Most of the disturbances are enclosed.
2. The second largest time constant from a self-regulating process is enclosed.
3. The largest time constant from a non-self-regulating process is enclosed.
4. The smallest process dead time is enclosed.
5. The measurement rangeability is sufficient.
6. The measurement accuracy is sufficient.
7. The measurement speed of response is sufficient.

The inner measurement error should be much less than the allowable control error of the outer loop divided by the process gain of the outer loop. The inner measurement dead time and time constant should be much less than the dead time of the inner loop process (McMillan, 1982).

The inner and outer controllers must be properly tuned to fully achieve the benefits of cascade control. The inner controller should be tuned first. The inner loop response should be as underdamped, as "hot" as possible, to minimize its period. Consequently proportional or proportional plus derivative control is preferable to proportional plus reset control because reset increases the loop period (offset of the inner controlled variable is usually unimportant). Inner loop noise and derivative action on set point changes may necessitate a larger proportional band, the addition of reset action, and a smaller derivative time. The most common type of inner loop is the flow loop where measurement noise necessitates the use of a proportional plus reset controller tuned for an overdamped, "sluggish" response. If the outer loop has a self-regulating process and the inner to outer dead time and time constant ratio is not extremely small, the outer controller should be detuned to prevent sustained oscillations. The following equation estimates how much wider the outer proportional band must be set.

$$PB_o' = PB_o{*}\left(A{*}\frac{TC_i}{TC_o} + B{*}\frac{TD_i}{TD_o} + C\right) \tag{11.4}$$

where

$PB_o'$ = widened outer loop proportional band
$PB_o$ = original outer loop proportional band
$A$ = 0.4 for self-regulating inner loop
= 1.0 for non-self-regulating inner loop
$B$ = 0.4 for self-regulating inner loop
= 0.0 for non-self-regulating inner loop
$C$ = 0.8 for self-regulating inner loop
= 2.0 for non-self-regulating inner loop
$TC_i$ = inner loop time constant
$TC_o$ = outer loop time constant
$TD_i$ = inner loop dead time
$TD_o$ = outer loop dead time

This detuning requirement of the outer controller arises because the ultimate period of the inner loop approaches that of the outer loop for an inner self-regulating process and exceeds the outer loop period for an inner non-self-regulating process. This detuning has not been quantitatively identified and has caused extensive problems in the commissioning of such loops. It is probably the source of the well-known rule that cascade control should be used only when the inner loop is faster than the outer loop.

Although cascade control appears simple in concept, the balancing requirement between the outer controller output and inner controller remote set point, the use of deviation display instead of dedicated remote set point display, derivative mode action on set point changes, poor rangeability of the inner controller measurement, and the detuning requirement of the outer controller for a slower inner loop have caused many cascade controllers to be left in manual. The first three problems are solved by options available in microprocessor-based controllers in distributed control systems. The rangeability problem can be prevented by checking the rangeability of the inner measurement compared with the installed rangeability of the control valve (an orifice meter can seldom be used for cascade control of pH to reagent flow). The last problem can be solved by using Equation 11.4 as a guide for increasing the proportional band of the outer controller.

## *11.2 Feedforward Control*

Feedforward control improves loop performance by measuring the disturbance and calculating a signal that has an effect equal to and opposite that of the disturbance on the controlled variable. Ratio flow control is a simple example of feedforward control. If the disturbance affects the controlled variable sooner than the calculated signal, then a lead time should be applied to the feedforward signal (a pure lead cannot be used because of noise). If the disturbance affects the controlled variable later than the calculated signal, then a time constant should be applied to the feedforward signal. Several time constants in series would be necessary to compensate for disturbance dead time if an analog module were used, since analog modules are not capable of simulating pure dead time. These analog modules are called dynamic compensators or lead–lag modules and are used extensively in boiler control systems. The more powerful microprocessor-based controllers can do digital dead time compensation of the feedforward signal. If the feedforward signal arrives too soon, it can cause inverse response. In general, it is good practice to set the feedforward gain and lead time to undercompensate slightly.

Figure 11.7 shows a block diagram and the method of estimating the feedforward signal gain, time constant, and dead time. The objective is to ensure that the feedforward signal arrives at the summing junction at the process input at the same time as the feedforward signal. The feedforward calculation gain ($K_{fc}$) is set equal to the load disturbance gain ($K_l$) divided by the feedforward measurement gain ($K_{fm}$). The feedforward calculation time constant ($TC_{fc}$) is set equal to the load disturbance time constant ($TC_l$), if this time constant is larger than the control valve time constant ($TC_v$) and the feedforward measurement time constant ($TC_{fm}$). If it is not larger, a lead time set equal to the largest of the control valve and feedforward measurement time constants should be used. Note that if disturbance enters at a midpoint in the process, any process time constant between

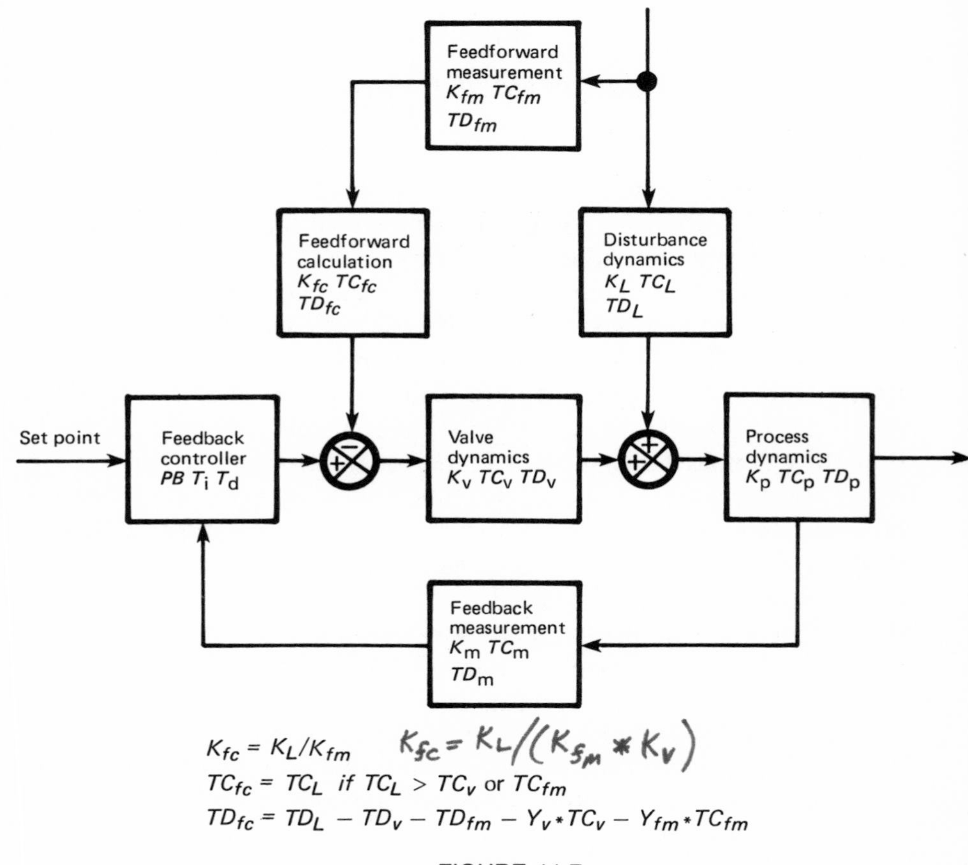


FIGURE 11.7
Block diagram of feedforward control loop.

the control valve and the disturbance can be treated as a control valve time constant. The feedforward calculation dead time ($TD_{fc}$) is set equal to the load disturbance dead time ($TD_l$) minus the sum of the control valve dead time ($TD_v$), the feedforward measurement dead time ($TD_{fm}$), the equivalent dead time of the control valve time constant ($Y_v$*$TC_v$), and the equivalent dead time of the feedforward measurement time constant ($Y_{fm}$*$TC_{fm}$). These equivalent dead times use $Y$ factors calculated by Equation 4.1 or read from Figure 4.2, if the feedforward calculation time constant is the largest time constant in the feedforward signal path.

The feedforward signal is shown being added to the controller output signal. In many cases, a multiplier is used instead of a summer. To help decide whether a multiplier or summer should be used, the required control valve signal should be plotted versus the feedforward measurement for several sets of operating conditions. If the slope of the plots varies the most, normally a multiplier is best so that the controller output can correct for this slope change. If the intercept of the plots varies the most, normally a summer is best so that the controller output can correct for this bias change. If the control valve signal should always go to zero when the feedforward measurement goes to zero, it is important to use a multiplier (multiplication by zero is always zero). Such is the case with flow feedforward for pH control. When the influent flow stops, the reagent flow should stop and not wait for the feedback measurement and the controller output correction. Multipliers are used for air-to-fuel ratio control because it is more important that the combustion control system correct for slope changes so that fuel-rich mixtures do not occur during load swings. Summers are used for three-element drum level control (steam flow is the feedforward signal to set the drum feedwater control valve) because the intercept varies due to blowdown flow changes and bias errors in the flow measurement signals. The summer that adds the level controller output signal to the steam flow signal typically has a −50 percent signal also added so that when the level controller output is +50 percent, the total correction to the steam flow signal is zero. The addition of a negative signal to the summer allows positive and negative corrections by the controller output to the feedforward signal.

Feedforward control reduces the open loop error seen by the feedback loop to the accuracy of the feedforward compensation. If the feedforward compensation accuracy is 10 percent, the open loop error, and consequently the peak and accumulated errors, are reduced by a factor of 10. The accuracy of the feedforward signal depends on the accuracy of the disturbance measurement, the accuracy of the equation that relates the disturbance to the controlled variable, and the accuracy of the arithmetic and dynamic computations.

## *11.3*
## *Signal Characterization*

Signal characterization improves loop performance by linearization of either the measurement or the valve gain. It generally requires the use of polynomial functions to fit an output versus input curve (computer programs can calculate the polynomial coefficients by linear regression). The more powerful microprocessor controllers have a sufficient variety of mathematical functional blocks and calculation steps easily to implement either input or output signal characterization. If the measurement or valve gain nonlinearity is predictable, the improvement in loop performance can be significant [see Chapter 9, Example 9.4, and McMillan (1980)]. The use of measurement signal characterization for pH control can eliminate the need of one well-mixed tank in a system that has several mixed tanks in series. Polynomial and logarithmic functions are used to calculate the *X* coordinate (ratio of reagent to influent flow) based on the *Y* coordinate (pH measurement) of the titration curve. The output of the controller is multiplied by the influent flow. The result is a linear reagent demand controller with built-in flow feedforward. The nonlinearity of the process due to the titration curve and influent flow (the process gain is inversely proportional to flow) is eliminated. If the titration curve is constant and the fit of the mathematical function is good in the control region, the improvement in control is dramatic.

If the steady-state plot of output versus input for any component in the loop is a curve whose shape is constant, signal characterization can be used. If the curve output is the manipulated variable, the controller output is the signal characterized. If the curve output is the controlled variable, the controller input (measurement) signal is the signal characterized (Shinskey, 1962). In each case, signal characterization applies the inverse of the gain by using an equation of the curve to calculate the *X*-axis coordinates from the *Y*-axis coordinates. The *Y*-axis is the original signal and the *X*-axis is the characterized signal.

## *11.4* *Mode Characterization*

Mode characterization improves loop performance by adjusting the controller mode settings for changing gains, time constants, and dead times. The notch gain or nonlinear controller is the most familiar example of mode characterization. The nonlinear controller's flexibility is limited to selection of a single notch gain and location. It does not compensate for continuous gain changes or for signal characterization. The titration curve's "S" shape may appear to be compensated graphically by the nonlinear controller's notch gain. However, the slope of the titration curve, which is the process gain, is continuously changing. In the extreme case of a strong acid and strong base, the slope changes by a factor of 10 for each pH unit deviation from neutrality. Figure 11.8b is a blown-up section of the apparently vertical slope near neutrality in Figure 11.8a. Notice that the blown-up section reveals another S-shaped titration curve.

The error squared controller provides a reduction in mode action to compensate for loop gains that decrease with error or to reduce control action when the measurement is near set point.

Microprocessor controllers will facilitate the adjustment of the controller mode settings based on calculated dead times and time constants (see Chapter 9). The momentary increase in derivative action by a factor of 10 can significantly reduce the peak error if it is exponentially decreased after the peak to its original value (McMillan, 1980). The reduction in peak error is greatest for characterized input signals in the region of high measurement gain (e.g., neutral region for pH).

The preload or batch controller can significantly reduce the overshoot of temperature during the startup of exothermic batch reactions. One such controller limits its output to a linear function of the error when the error is within the percentage specified on the bias dial setting and the controller output is saturated. If the proportional-integral-derivative (PID) output is greater than this limit, the output ramps linearly with error (no pro-

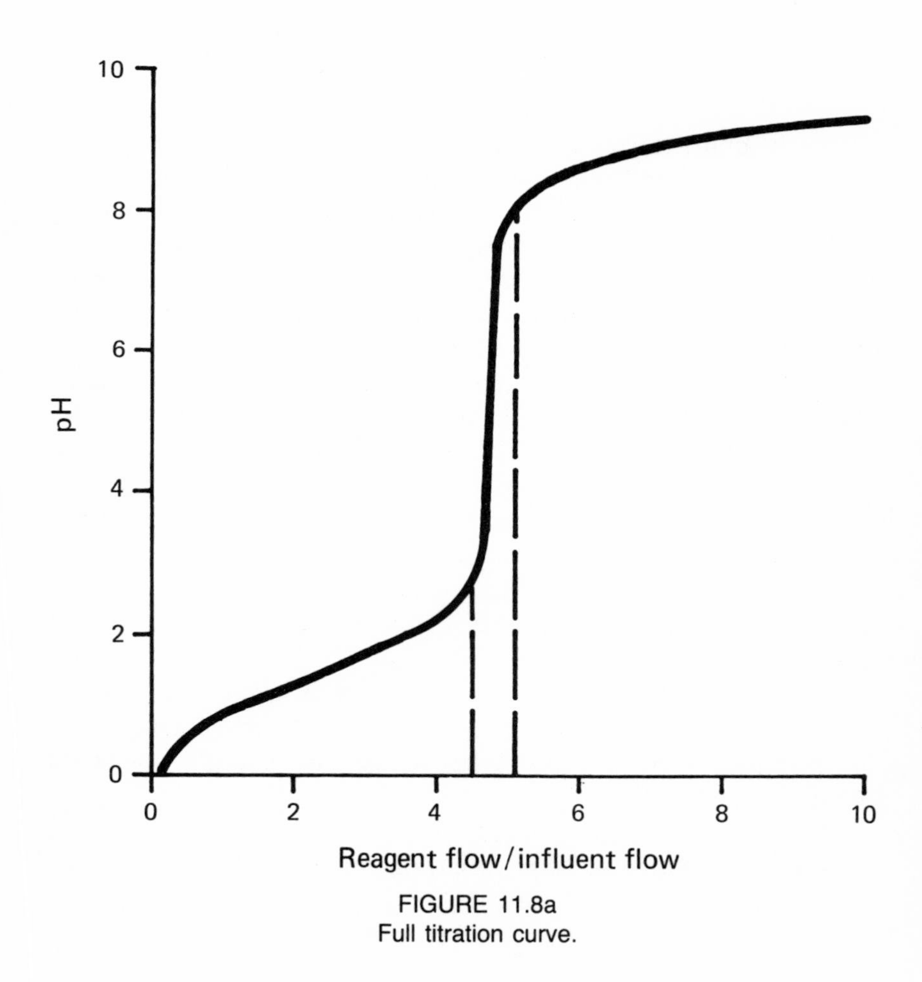


FIGURE 11.8a
Full titration curve.

portional, reset, or rate action). If the PID output becomes less than the limit, normal PID action resumes. If the switch is in the "auto" position on the controller side panel, batch action will take over whenever the PID output saturates and the error is within the percentage specified on the bias dial setting. If the mode switch is in the "semi auto" position, the mode switch must be turned to "off" and then back to "semi auto" or "auto" to rearm the controller for batch action. A further improvement could be made by compensating the controller modes for the reduction in the heat of reaction and the increase in the heat transfer area near the end of the batch.

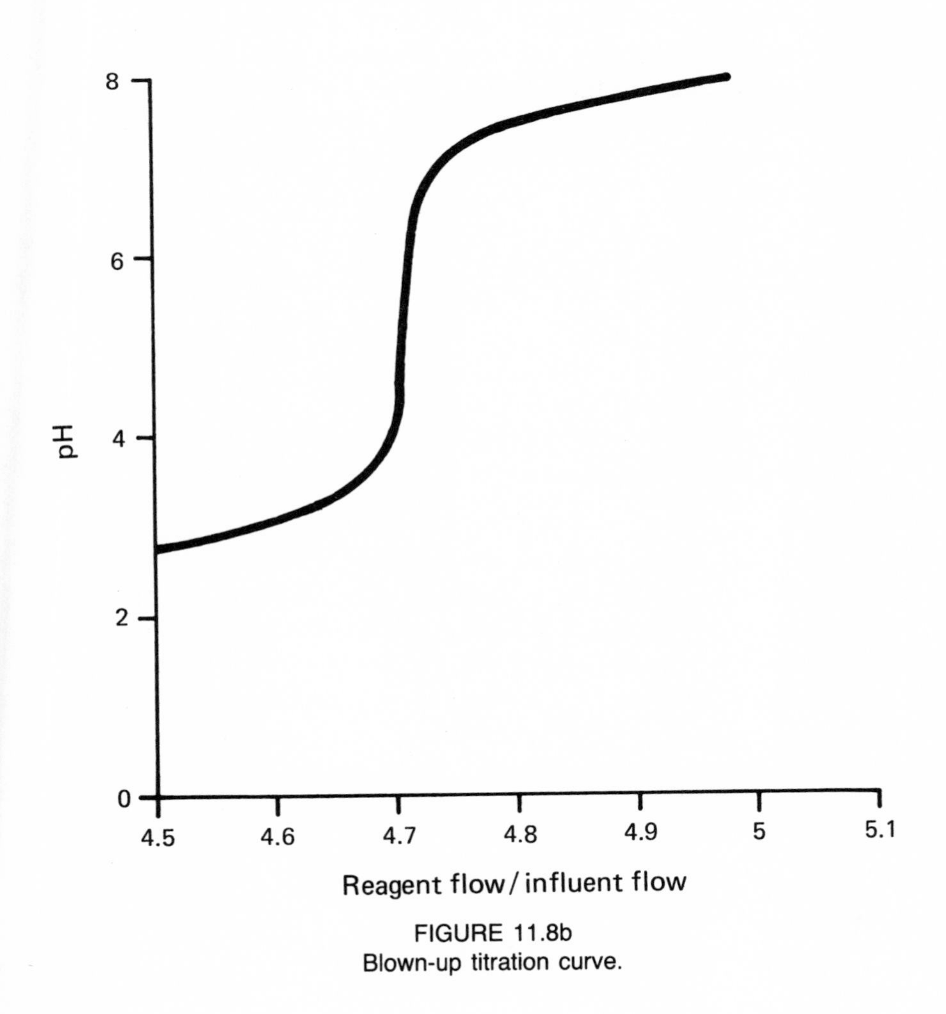


FIGURE 11.8b
Blown-up titration curve.

## 11.5 Direct Synthesis Controller

A digital controller algorithm can be tailor-made to provide a closed loop response to set point changes that has no overshoot or oscillation if the loop has a fixed time constant, dead time, and linear gain. The direct synthesis controller algorithm, also known as the Dahlin algorithm, contains a complementary model of the process to provide a desired response to step set point changes. The desired closed loop response is described by a selected dead time and time constant. This selected dead time must be close to the actual dead time of the loop to prevent instability. For a loop with pure dead time (no time constant), the selected dead time must satisfy the following inequality (Badavas, 1981):

$$0.5*TD_o < TD_c < 1.25*TD_o \quad (11.5)$$

where

$TD_o$ = dead time of the open loop response
$TD_c$ = dead time selected for the closed loop response

The size of this window of selectable dead times becomes larger for loops with a time constant. The larger the time constant, the larger is the window. A difference equation for the programming of a direct synthesis controller for a loop with a first-order plus dead-time self-regulating process is as follows (see Appendix D for the development of Equation 11.6):

$$O_n = B*O_{n-1} + (1 - B)*O_{n-N-1} + K'*(E_n - A*E_{n-1}) \quad (11.6)$$

$$A = e^{-T_s/TC_o} \quad (11.7)$$

$$B = e^{-T_s/TC_c} \quad (11.8)$$

where

$O$ = controller output
$E$ = controller error (set point minus measurement)
$TC_o$ = time constant of the open loop response
$TC_c$ = time constant selected for the closed loop response
$T_s$ = sample time of digital controller
$n$ = subscript denoting present value sampled
$n - 1$ = subscript denoting value from one sample period ago
$n - N$ = subscript denoting value from $N$ sample periods ago
$K'$ = gain factor

If the time constant selected for the closed loop response is set very small, the measurement will track a set point change closely, but will be delayed by the selected dead time. However the signal to the control valve alternately will decrease and increase by large amounts from one sample period to another. This phenomenon, known as ringing, is undesirable because the control valve may not be able to track such rapid changes or the control valve and its accessories may wear out prematurely. This ringing can be eliminated by modification of the digital algorithm if the open loop dead time is small compared with the open loop time constant. The resulting closed loop response can become oscillatory for set point changes if small values of the closed loop time constant are selected. This modified direct synthesis controller is equivalent to a proportional plus integral (PI) controller for specific mode settings (Corripio, 1973). For load disturbances to processes with a small open loop dead time compared with the open loop time constant, the direct synthesis controller does not perform better than a properly tuned conventional three-mode (PID) controller.

The direct synthesis controller is equivalent to a dead-beat controller if the closed loop response time constant is set to zero. The direct synthesis controller is identical to the Smith Predictor dead-time compensator if the closed loop dead time is set equal to the open loop dead time and the closed loop time constant is set equal to the open loop time constant.

## *11.6*
## *Dead-Time Compensation*

Since dead time is the cause of poor loop performance, there have been various attempts to cancel its effect in the control loop. The Smith Predictor cancels the dead-time term from the transfer function of the loop for set point changes by adding the controller output signal after it has passed through a filter and a pulse function. The pulse function consists of the delayed filter output subtracted from the immediate filter output. Figure 11.9 is a block diagram of a Smith Predictor installed on the output of a conventional controller for a self-regulating process. The filter time constant should be set equal to the process time constant and the pulse function time delay should be set equal to the process dead time. The Smith Predictor does not cancel out dead time from the loop transfer function for load disturbances. If the dead time were truly canceled out, leaving just a single process time constant, the proportional band could be set to zero (infinite gain) without the controller going unstable. In real applications, a single time constant does not exist, the dead time is not known exactly, and load disturbances occur.

The Smith Predictor has been heavily touted in the literature for its effectiveness as compared with conventional controllers. But most of these comparisons were made with proportional plus integral (PI) controllers. A dynamic simulation of a conventional three-mode (PID) controller tuned per Chapter 4 equations and a Smith Predictor for dead-time to time-constant ratios ranging from 0.1 to 10 showed that the Smith Predictor performance was not as good for self-regulating processes and potentially dangerous for non-self-regulating processes. The closed loop responses for the two controllers are plotted in Figures 11.10–11.12. Figure 11.10 shows that the Smith Predictor will cause a slightly larger peak error (10 percent typically) and a significantly larger accumulated error (100 percent typically) than a conventional three-mode (PID) controller for disturbances to self-regulating processes with dead-time to time-constant ratios of less than 10. Figures 11.11 and 11.12 show that the Smith Predictor develops a large offset for integrating

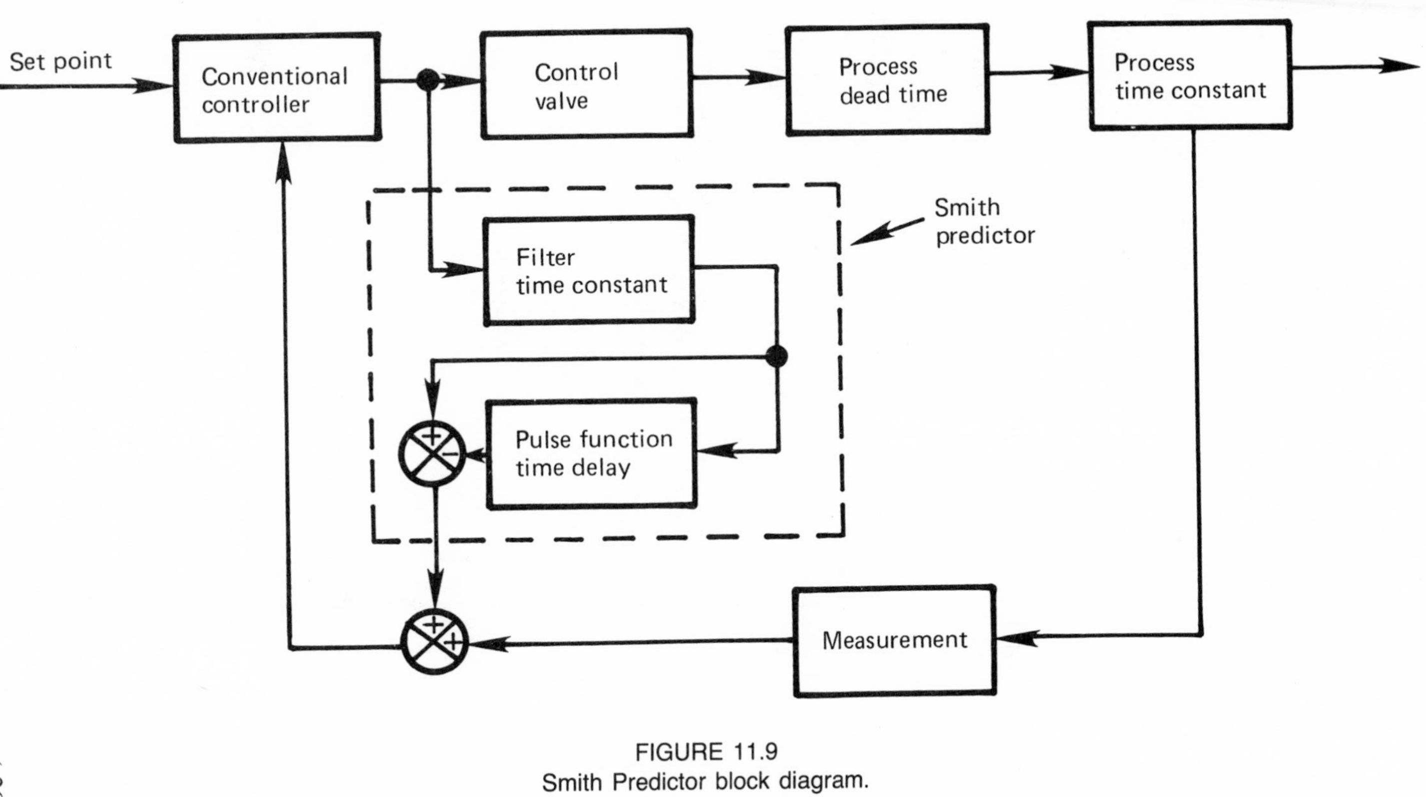


FIGURE 11.9
Smith Predictor block diagram.

processes (the effect of reset action is canceled) and ramps off to a physical limit for runaway processes (the window of allowable proportional bands is closed).

For non-self-regulating processes, another dynamic block was added to the output of the filter in the Smith Predictor that had an integrator gain or runaway time constant equal to that in the process. If the proportional band is decreased in an attempt to improve performance, the Smith Predictor will amplify what might be negligible noise levels to the point where they deteriorate loop performance. The Smith Predictor will be more effective than a conventional PID controller for set point rather than load changes. For very large dead-time to time-constant ratios experienced in chromatograph loops, the more gradual overdamped response of the Smith Predictor may be desirable. Also, the dead time due to analysis sample time is predictable and relatively constant so that the delay time in the pulse function can be set accurately. Most of the industrial applications of Smith Predictors have been in loops with on-line or laboratory analysis (laboratory analysis is entered by the operator as measurement data periodically to emulate a sample and hold analysis signal). Figure 11.13 shows the use of a Smith Predictor for carbon monoxide analysis control of flue gas.

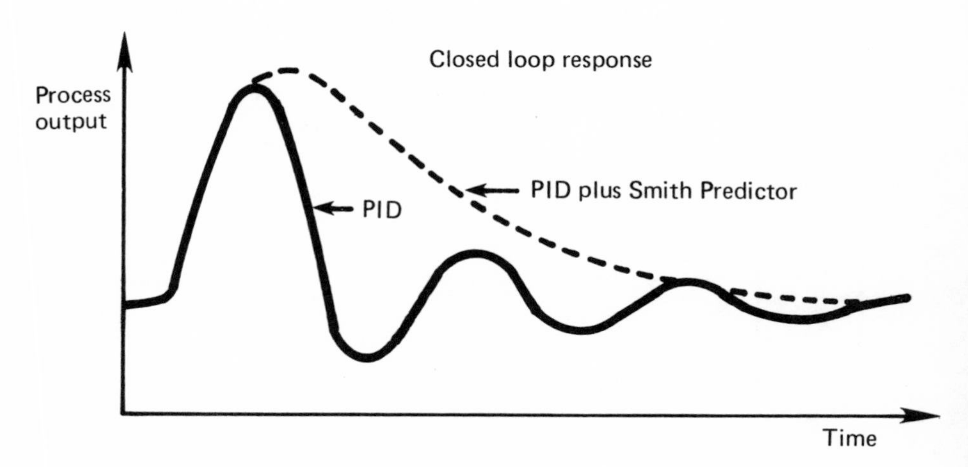


FIGURE 11.10
Comparison of Smith Predictor response with conventional PID controller (self-regulating process).

# *11.7*
# *Self-Tuning Controllers and Regulators*

The two major classes are self-tuning deterministic controllers and stochastic regulators. The self-tuning deterministic controllers identify the loop gain and dynamics after initiating deliberate upsets in the controller set point or output. The identified loop model is then used to calculate the controller mode settings. The computational capability and flexibility of microprocessor-based controllers have led to the packaging and marketing of these self-tuning controllers by several manufacturers. Most of these self-tuning controllers currently available use set point changes and a conventional three-mode controller. The operator is typically given the option of selecting the size of the set point upset and checking the calculated mode settings before they are used to update the existing mode settings. One particular controller provides self-diagnostics for troubleshooting and a safe mode of operation for power failures (Andreiev, 1973). These controllers do not recognize the possibility of a proportional band window for stability and thus should not be

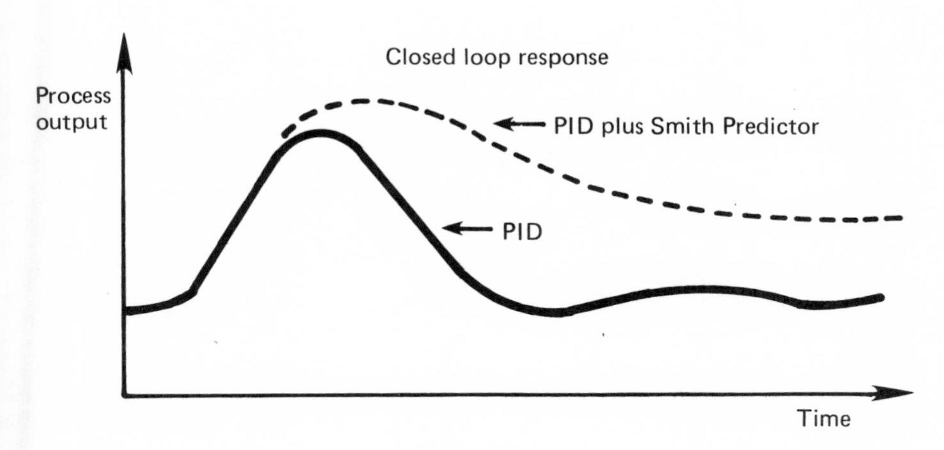


FIGURE 11.11
Comparison of Smith Predictor response with conventional PID controller (integrating process).

used on runaway processes without extensive prior on-line testing. If the size of the permissible set point upset is small because of operational constraints, the set-point to noise ratio may be too small for adequate identification.

Self-tuning stochastic regulators are designed to work on loops with low signal to noise ratios. These are termed regulators instead of controllers because they are not designed specifically for set point changes. No set point or output upsets are applied to the loop. The existing disturbances during closed loop operation are used to estimate a set of parameters to minimize the mean square deviation from set point that provides minimum variance of the deviation. If large and rapid swings of the control valve are undesirable, the regulator can provide a combination of minimum variance of deviation and control valve movement. Note that despite some previous claims to the contrary in the literature, identification from just existing closed loop disturbances yields only the controller gain and dynamics and not those of the other loop components. The self-tuning stochastic regulator does not claim to identify the loop components. K. J. Astrom has successfully tested the software on a paper machine, digester, ore crusher, enthalpy exchanger, and supertanker. The user must determine in advance the number of parameters in the estimator, the initial values of the param-

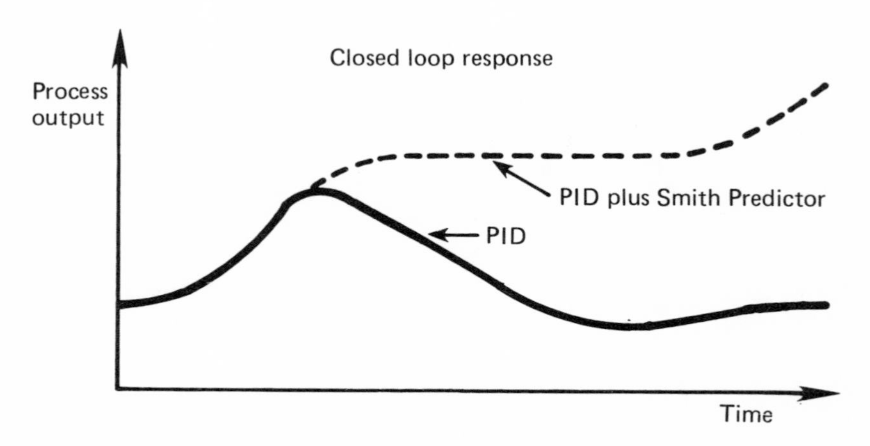


FIGURE 11.12
Comparison of Smith Predictor response with conventional PID controller (runaway process).

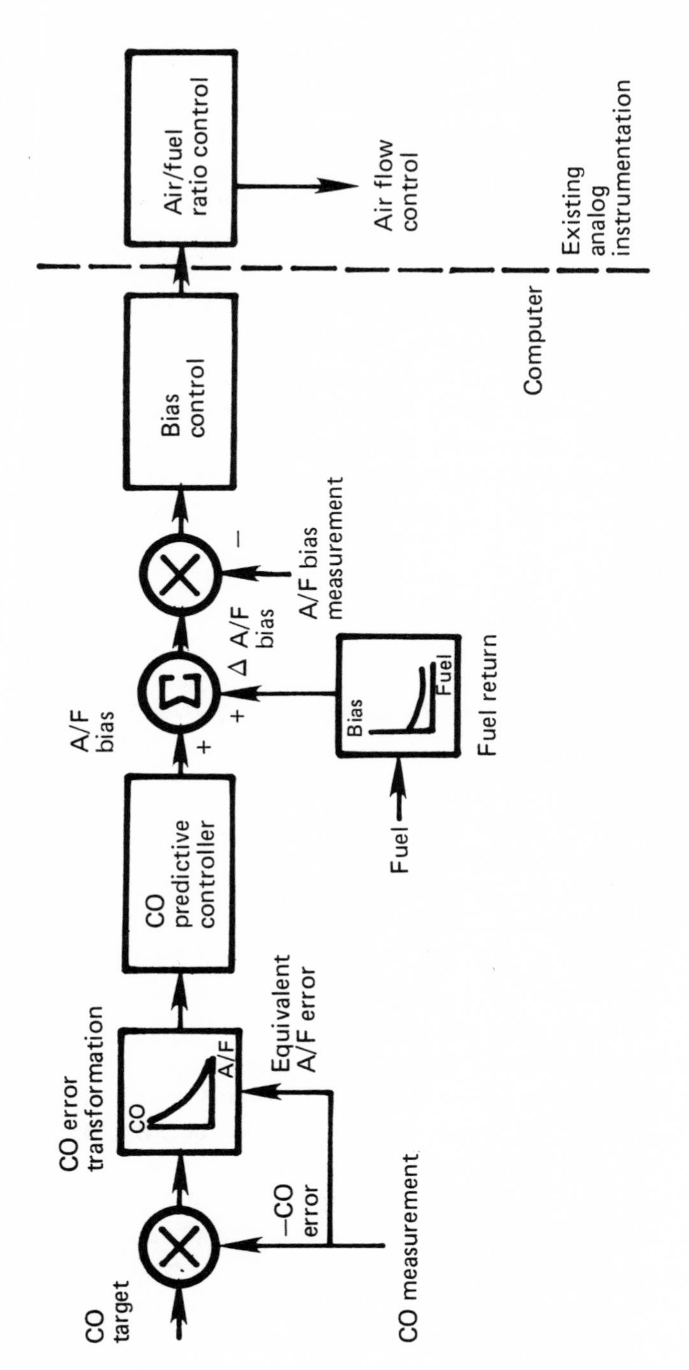


FIGURE 11.13
Smith Predictor for flue gas CO control.

eters, the rate of exponential forgetting, and the sample rate (Astrom et al., 1977). The initial values selected may not be critical but the rate of exponential forgetting should be carefully set on the basis of how frequent the changes in loop characteristics are and how persistent the disturbances are. If the loop becomes quiet for long periods of time, the estimated parameters may be forgotten. The success of the regulator depends upon sufficient disturbances for estimation of the parameters. If the regulator is doing an excellent job of minimizing the effect of the disturbances, or the size or number of disturbances is small, the resulting mismatch of parameters can cause a temporary positive feedback condition. One recently developed software package is claimed to have proprietary features to solve this problem (Fjeld & Wilhelm, 1981). This software package has been successfully applied to control moisture content in a paper plant.

## *11.8*
## *Model Predictive Controller*

If a controller can be designed to be sufficiently robust (tolerant of changes in loop gain and dynamics), the loop characteristics need only be identified infrequently to update the controller. The model predictive controller developed by J. Richalet appears to be robust enough for many industrial applications. The controller identifies loop characteristics from an open loop test. This test is performed initially and typically yearly or after major equipment, operational, or environmental changes. The characteristics are used to develop a trajectory of future loop response if there were no further control actions or disturbances. A desired reference trajectory is also computed to prevent the closed loop response from being too "hot" to avoid excessive control valve movement. A single parameter is used to set this response. A selected value of this parameter gives the same results from one loop to another independent of loop characteristics. The future loop response for no new disturbances or control actions is modified to predict the effect of control actions. The resulting loop response trajectory is compared with the reference trajectory to generate a vector of errors and future control actions (Richalet, 1978). The exact method of computing the vector of future control actions from the vector of errors is not disclosed. Richalet has successfully applied the model predictive controller to a distillation column, a complete PVC plant, and a steam generator in a power plant.

A similar strategy, called dynamic matrix control, has been independently developed by C. R. Cutler and B. L. Ramaker of Shell Oil Company. The dynamic matrix controller uses a penalty for excessive valve motion instead of a reference trajectory to prevent too hot a closed loop response.

The valve motion penalty also provides a single paramater independent of loop characteristics to set the closed loop response. The computation of the controller output requires an on-line matrix inversion. The entire set of required computations is well documented. The dynamic matrix controller has been in service in various applications in Shell since 1973 (Cutler & Ramaker, 1980).

The model predictive and dynamic matrix controllers provide a single tuning parameter independent of the loop, provide the operator with a trajectory of future loop response for manual control and for reassurance during automatic control, and facilitate the easy implementation of feedforward control for measurable disturbances. These controllers require much more computation and memory than conventional three-mode controllers but such capability is already available in the more powerful microprocessor-based controllers in distributed control systems. Since an open loop response is employed, these controllers should not be used for runaway processes or processes with ringing (open loop oscillations) without extensive prior on-line testing. These controllers may not be suitable as the outer controller in a cascade loop because the outer loop controller sees extensive ringing for a tightly tuned inner loop. Also since these controllers like the Smith Predictor and Direct Synthesis Controller are designed to minimize overshoot for set point changes, the errors for load disturbances are larger than those for a well tuned PID controller.

# CHAPTER XII
# *Summary*

To maximize loop performance (see Figure 12.1):

1. Minimize dead time wherever it appears in the loop.
2. Minimize all instrument time constants.
3. Maximize the largest negative feedback time constant in a self-regulating process.
4. Minimize all negative feedback time constants smaller than the largest in a self-regulating process.
5. Minimize all negative feedback time constants in a non-self-regulating process.
6. Maximize the positive feedback time constant in a non-self-regulating process.
7. Minimize the process, integrator, and disturbance gains.
8. Maximize the disturbance time constant and time interval.
9. Minimize the proportional band and integral time setting of the controller.
10. Maximize the derivative time setting of the controller.

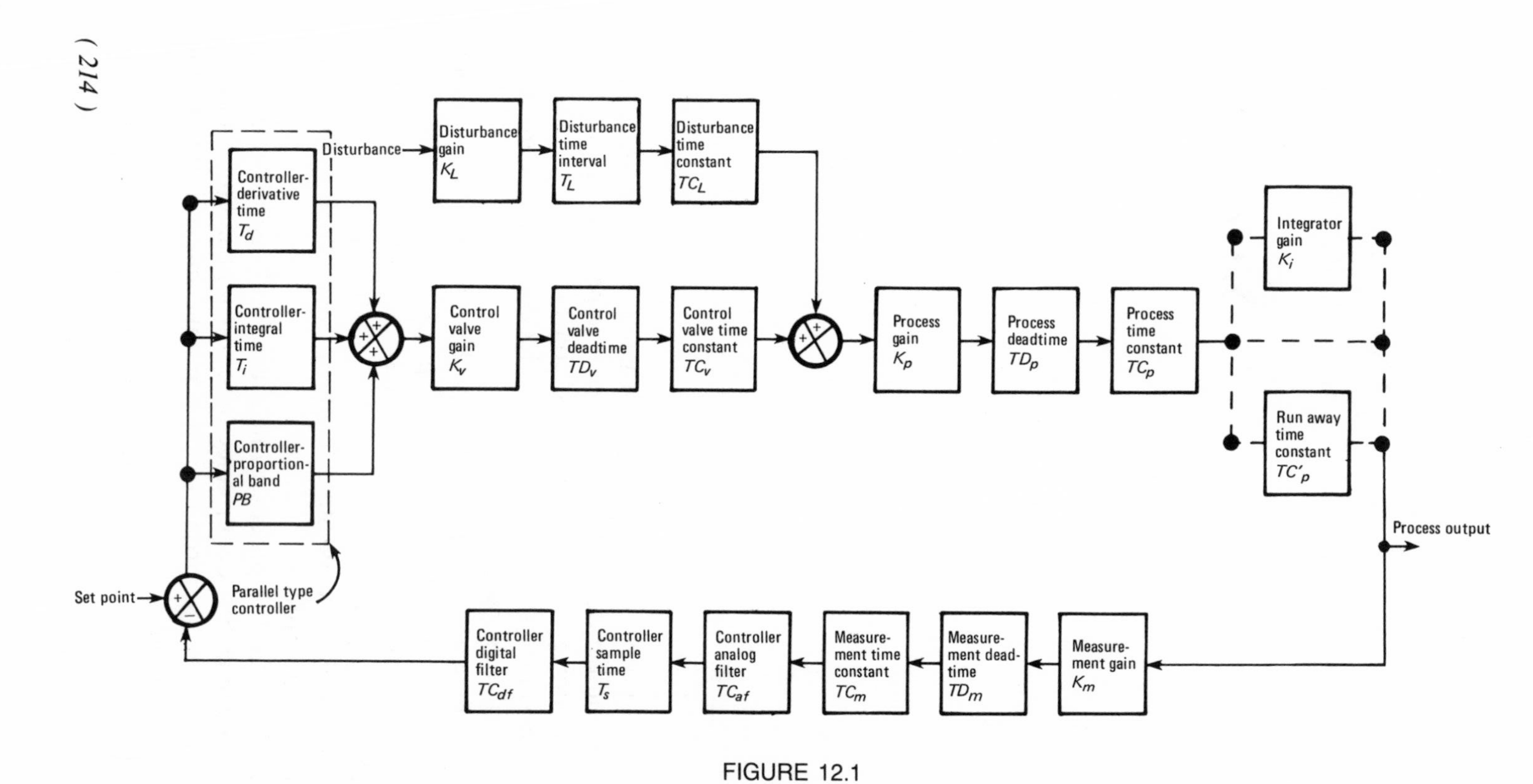


FIGURE 12.1
Block diagram of loop components.

To estimate the mode settings and loop performance:

1. Measure by open loop test, or estimate by the use of the equations in Chapters 4–7, the dead times, time constants, and gains of the process, disturbance, and instruments. If the measurement is noisy, estimate the filter time constant by Equation 6.2 or the proportional band by Equation 8.3 required for attenuation.

2. Convert each of the time constants smaller than the largest negative feedback time constant in the loop to equivalent dead time by Equation 4.1 or Figure 4.2.

3. Sum up all the equivalent dead times in the loop.

4. Sum up all the negative feedback time constants in the loop and subtract from this total the sum of the equivalent dead times.

5. Sum up all the pure dead times in the loop and add to this total the sum of equivalent dead times.

6. Use the total loop negative feedback time constant from step 4 and the total loop dead time from step 5 in Equation 4.5 for self-regulating processes, in Equation 4.15 for integrating processes, and in Equation 4.21 for runaway processes to estimate the ultimate period.

7. Use the total loop negative feedback time constant from step 4 and the total loop dead time from step 5 in Equation 4.11 for self-regulating processes, in Equation 4.16 for integrating processes, and in Equation 4.22 for runaway processes to estimate the proportional band. Use the proportional band setting from step 1 for noise attenuation, if it is larger. For runaway processes, divide the maximum proportional band from Equation 4.26 by one half of the proportional band from Equation 4.22 to estimate the size of the proportional band window.

8. Use the ultimate period from step 6 in Equation 4.7 for a PI controller or in Equation 4.9 for a PID controller to estimate the integral time.

9. Choose the controller mode settings available that are closest to the estimated values. Remember to invert the integral time if the integral mode setting is in terms of repeats per minute.

10. Estimate the peak error by Equation 3.10 and the accumulated error by Equation 3.4.† If the largest instrument time

† If the measurement is not noisy and just the peak and accumulated error as a fraction of the open loop error needs to be estimated, equations 4.13, 4.14, 4.19, 4.20, 4.24, and 4.25 can be used, which do not require knowledge of the loop steady state gains.

constant was larger than the largest negative feedback time constant in the process, multiply the results by the ratio of this instrument time constant to the process time constant.

11. If the errors are too large, investigate the feasibility of cascade control for isolatable disturbances or feedforward control for measurable disturbances.

# References

Anderson, G. D., "Basics of Level Control," *Instruments and Control Systems,* vol. 52, no. 10, 1979, pp. 33–36.

Andreiev, N., "A New Dimension: A Self-Tuning Controller That Optimizes PID Constants," *Control Engineering,* Aug. 1981, pp. 84–85.

Astrom, K. J., et al., "Theory and Applications of Self-Tuning Regulators," *Automatica,* vol. 13, 1977, pp. 457–476.

Bailey, J. E., and Ollis, D. F., *Biochemical Engineering Fundamentals,* McGraw-Hill, New York, 1977.

Baumann, H. D., "Control Valve with Laminar Flow Characteristic," *Advances in Instrumentation,* ISA Conference, St. Louis, vol. 36, part 1, paper 540, March 1981.

Badavas, P. C., "Stability Limits and Parameter Sensitivity in Synthesized Digital Controllers," *Control Engineering,* Sept. 1981, pp. 120–124.

Bergeron, L., *Water Hammer in Hydraulics,* Wiley, New York, 1961, pp. 2–10.

Bor, T., "The Static Mixer as a Chemical Reactor," *British Chemical Engineering,* vol. 16, no. 7, 1971, p. 610.

Bradner, M., "Pneumatic Transmission Lag," *Instruments,* vol. 22, July 1949, pp. 618–621.

Brez, C., "Predicting Limit Cycling in Control Valves," *Control Engineering,* vol. 22, no. 7, 1975, pp. 35–37.

Bristol, E. H., "Adaptive Control Odyssey," *Advances in Instrumentation,* ISA 25th Annual Conference, Philadelphia, vol. 25, part 1, paper 561, Oct. 1970.

Bristol, E. H., "Designing and Programming Control Algorithms for DDC Systems," *Control Engineering,* vol. 24, no. 1, 1977, pp. 24–26.

Buchwald, W. G., "Control Performance with Reduced Valve Pressure Drop for Energy Conservation," *Advances in Instrumen-*

*tation,* ISA International Instrumentation/Automation Conference, New York, vol. 29, part 2, paper 610, Oct. 1974.

Buckley, P. S., "Dynamics of Temperature Measurement," *Process Annual Symposium,* Instrumentation Process Industry, Wilmington, vol. 34, 1979, pp. 21–29.

Carroll, R. M., et al., "Studies of Sheathed Thermocouple Construction and Installation in Thermowells to Obtain Faster Response," *Sixth International Symposium on Temperature,* paper 19-4, 1982.

Corripio, A. B., "Digital Control Algorithms—Part 1. Dahlin Algorithm," *Instruments and Control Systems,* Oct. 1973, pp. 57–59.

Cutler, C. R., and Ramaker, B. L., "Dynamic Matrix Control—A Computer Control Algorithm," *JACC Proceedings,* San Francisco, paper WP5-B, 1980.

Fertik, H. A., "Tuning Controllers for Noisy Processes," *ISA Transactions,* vol. 14, no. 4, 1975, p. 292.

Fjeld, M., and Wilhelm, R. G., "Self-Tuning Regulators—The Software Way," *Control Engineering,* Oct. 1981, pp. 99–102.

Giusti, A. L., Jr., and Hougen, J. O., "Dynamics of pH Electrodes," *Control Engineering,* vol. 8, no. 4, 1961, p. 136.

Goff, K. W., "Estimating Characteristics and Effects of Noisy Signals," *ISA Journal,* vol. 13, no. 11, 1966, p. 45.

Goff, K. W., "A Systematic Approach to DDC Design," *ISA Journal,* vol. 13, no. 12, 1966, p. 44.

Green, R. M., and Field, W. B., "Evaluation of pH Measurement Response," *AID Symposium Proceedings,* San Francisco, June 1964, pp. 49–58.

Harriot, P., *Process Control,* McGraw-Hill, New York, 1964, pp. 100–102, 220–222, 258–266, 312–314.

Heider, R. L., "How Digital Controllers Can Cause Process Cycling," *ACC Conference Proceedings,* Arlington, Va., June 1982.

Kardos, P. W., "Response of Temperature Measuring Elements," *Chemical Engineering,* Aug. 29, 1977, pp. 79–83.

Kerlin, T. W., et al., "Time Response of Installed Temperature Sensors," *Sixth International Symposium on Temperature,* paper 19-2, 1982.

Liptak, B. G., *Instrument Engineers Handbook, Volume I. Process Measurement,* Chilton, 1969, pp. 6–14.

Lloyd, S. G., "Guidelines for the Application of Valve Positioners," *Technical Monograph 23,* Fisher Controls Co., 1969.

Lloyd, S. G., and Anderson, G. D., *Industrial Process Control,* Fisher Controls Co., 1971, pp. 48–50 and pp. 222–224.

Loeber, C., "A Digital Filter with Adaptive Delay for Smoothing Stochastic Measurement Signals," *Measurement for Progress in Science and Industry,* 8th IMEKO Congress, 1980, pp. 325–329.

Luyben, W. L., and Cheung, T. F., "Proportional-Derivative Control of Exothermic Reactors," *Joint Automatic Control Conference,* Philadelphia, vol. 3, Oct. 1978, pp. 379–390.

Luyben, W. L., and Melcic, M., "Consider Reactor Control Lags," *Hydrocarbon Processing,* vol. 57, no. 3, 1978, pp. 115–117.

Mamzic, C. L., "Improving the Dynamics of Pneumatic Positioners," *ISA Journal,* vol. 5, no. 8, Aug. 1958, pp. 38–43.

McMillan, G. K., "Guidelines for the Measurement and Estimation of Electrical Conductivity for Chemical Solutions," *Advances in Instrumentation,* ISA Conference, Philadelphia, vol. 33, part 1, paper 705, Oct. 1978.

McMillan, G. K., "Process Simulation of a Waste Treatment Neutralization Process," *Instrumentation in the Chemical and Petroleum Industries,* New Orleans, *ISA Proceedings,* vol. 15, Apr. 1979, pp. 27–38.

McMillan, G. K., "Adaptive Control for a Waste Neutralization Process," *Instrumentation in the Chemical and Petroleum Industries,* New Orleans, *ISA Proceedings,* vol. 15, Apr. 1979, pp. 39–49.

McMillan, G. K., "Energy Savings by Improved Furnace Pressure Control," *Instrumentation in the Chemical and Petroleum Industries,* Secaucus, N.J., *ISA Proceedings,* vol. 16, April 1980, pp. 145–149.

McMillan, G. K., "Improved Control By Signal Linearization," *Advances in Instrumentation,* St. Louis, *ISA Conference Proceedings,* vol. 36, part 1, paper 521, March 1981 (a).

McMillan, G. K., "The Effect of Process Non-Self-Regulation on Control Loop Performance," *ISA Conference Proceedings,* Anaheim, Oct. 1981 (b).

McMillan, G. K., "The Effect of Cascade Control on Loop Performance," *ACC Conference Proceedings,* Arlington, June 1982.

Moore, C. F., Smith, C. L., and Murrill, P. W., "Simplifying Digital Control Dynamics for Controller Tuning and Hardware Lag Effects," *Instrument Practice,* Jan. 1969, pp. 45–49.

Morse, R. H., et al., "Aspects of Tuning a Boiler Control System—A Strategy for Optimization," *ISA Transactions,* vol. 16, no. 1, 1977, pp. 15–31.

Murrill, P. W., et al., "A Comparison of Controller Tuning Techniques," *Control Engineering,* Dec. 1967, pp. 72–75.

Myron, T. J., "Digital Technology in Process Control," *Computer Design,* Nov. 1981, pp. 117–120.

Schuder, C. B., "Control Valve Rangeability and the Use of Valve Positioners," *Advances in Instrumentation,* ISA 26th Annual Conference, Chicago, vol. 26, part 4, paper 817, Oct. 1971.

Shinskey, F. G., "Controls for Nonlinear Processes," *Chemical Engineering,* vol. 69, no. 6, 1962, pp. 155–158.

Shinskey, F. G., *pH and pION Control in Process and Waste Streams,* McGraw-Hill, New York, 1973, pp. 52–56.

Shinskey, F. G., *Energy Conservation Through Control,* McGraw-Hill, New York, 1978, pp. 31–36.

Shinskey, F. G., *Process Control Systems,* second edition, McGraw-Hill, New York, 1979, pp. 57–111, 228–230.

Sood, M., and Huddleston, H. T., "Tuning Controllers for Random Disturbances," *Instrumentation Technology,* vol. 24, no. 2, 1977, pp. 61–63.

Stout, T. M., "Control System Justification," *Instrumentation Technology,* vol. 23, no. 9, 1976, pp. 51–58.

Weiss, M. D., "Gas Chromatographs vs. Mass Spectrometers on Line," *Control Engineering,* Sept. 1977, pp. 66–68.

Weber, R., and Zumwalt, R. E., "The Effect of Measurement Noise on Feedback Controllers With Signal Selectors," *AIChE 86th National Meeting,* Houston, Apr. 1979.

Yekutiel, O., "A Reduced-Delay Sampled Data Hold," *JACC Proceedings,* San Francisco, paper WA4-E, 1980.

Ziegler, J. G., and Nichols, N. B., "Optimum Settings for Automatic Controllers," *ASME Transactions,* 1942, pp. 759–768.

Ziegler, J. G., and Nichols, N. B., "Process Lags in Automatic Control Circuits," *ASME Transactions,* 1943, pp. 433–444.

APPENDIX 

# Equation for the Accumulated Error

The equation for the accumulated error can be developed from the equation for the output of the parallel proportional plus reset controller.

$$C(T1) - C(T2) = \frac{100}{PB}*\left[E(T1) - E(T2) + \frac{1}{T_i}*\int E(T)*dT\right] \quad (A1)$$

where

$C(T1)$ = controller output at time $T1$
$C(T2)$ = controller output at time $T2$
$PB$ = proportional band of the controller (100 percent gain)
$E(T1)$ = error at time $T1$
$E(T2)$ = error at time $T2$
$T_i$ = reset time of the controller (minutes/repeat)

The error before the disturbance is zero and the error after the controller has completed the correction is also zero.

$(E(T1) = E(T2) = 0)$

Substitute $\Delta C = C(T1) - C(T2)$ and $E(T1) = E(T2) = 0$ into Equation A1:

$$\Delta C = \left[\frac{100}{PB^*T_i}\right]^* \int E(T)^*dT \qquad \text{(A2)}$$

Definition of accumulated error:

$$E_i = \int E(T)^*dT \qquad \text{(A3)}$$

Substitute Equation A3 into A2:

$$E_i = \left[\frac{PB^*T_i}{100}\right]^*\Delta C \qquad \text{(A4)}$$

Equation A4 is identical to Equation 3.7 in Chapter 3 with $K_m = 1$.

APPENDIX B

# Equation for the Peak Error

The minimum possible peak error for a self-regulating process for a given step disturbance and perfect controller response can be calculated if the process dead time and time constant are known. If the process dead time precedes the process time constant, the controller cannot affect the controlled variable until one dead time after the process starts to change exponentially. If the process time constant precedes the process dead time, the controller cannot sense the error until one dead time after the process starts to change exponentially. In either case, the peak error is the exponential response of the process to the step disturbance after one dead time.

$$E_x = \left[1 - e^{-[TD/TC]}\right] * E_o \qquad \text{(B1)}$$

where

$E_x$ = peak (maximum) error
$e$ = base of the natural logarithm ($e$ = 2.71828)
$TD$ = process dead time
$TC$ = process time constant

Approximate the exponential

$$e^{-[TD/TC]}$$

as

$$\left[\frac{1}{1 + \left[\frac{TD}{TC}\right]}\right]$$

for the small ratios of $\left[\frac{TD}{TC}\right]$ that are typical for chemical processes:

$$E_x = \left[1 - \frac{1}{1 + \left[\frac{TD}{TC}\right]}\right] {*} E_o \tag{B2}$$

Combine the terms over a common denominator:

$$E_x = \left[\frac{1 + \left[\frac{TD}{TC}\right] - 1}{1 + \left[\frac{TD}{TC}\right]}\right] {*} E_o \tag{B3}$$

Cancel out +1 and −1 in numerator of B3:

$$E_x = \left[\frac{\left[\frac{TD}{TC}\right]}{1 + \left[\frac{TD}{TC}\right]}\right] {*} E_o \tag{B4}$$

Multiply numerator and denominator by $K_e*TC$:

$$E_x = \left[\frac{K_e*TD}{TC + K_e*TD}\right]*E_o \tag{B5}$$

Simplify Equation B4 since $TD/TC < 1$ for most chemical processes:

$$E_x = \left[\frac{TD}{TC}\right]*E_o \tag{B6}$$

Substitute Equation 4.11 for *PB* of self-regulating process into B6, set $T_u = 4*TD$, and set $K_g = \frac{2\pi}{4}$:

$$E_x = \left[\frac{PB}{100*K_v*K_p*K_m}\right]*E_o \tag{B7}$$

Multiply the minimum possible error for perfect controller by 1.1 to get the actual peak error for quarter-amplitude damped response:

$$E_x = \left[\frac{1.1*PB}{100*K_v*K_p*K_m}\right]*E_o \tag{B8}$$

Equation B8 is identical to Equation 3.10 in Chapter 3 where $K = 1.1$.

APPENDIX C

# Equations for Mass and Energy Balance

The time constant, $TC$, and process gain, $K_p$, can be determined for blending, reaction (if reaction rate is fast and the reaction is not exothermic), level, and pressure control loops from a simple mass balance equation. The time constant, $TC$, and process gain, $K_p$, can be determined for some temperature and speed control loops from a simple energy balance equation. The general form of the simple material and energy balance equation is as follows, where the accumulation of mass or energy is equal to the integral of the flow into minus the flow out of mass or energy:

$$TC^*X_o = \int (K_p{}^*X_i - X_o)^*dT \tag{C1}$$

where

$X_o$ = mass or energy out
$X_i$ = mass or energy in

$TC$ = time constant
$K_p$ = process gain

The above equation with the $-X_o$ term describes a negative feedback or self-regulating process. If the flow of mass or energy out was zero or was constant so that there was no $-X_o$ term, then the equation describes a no-feedback or integrating process. If the minus sign was replaced with a positive sign to give a $+X_o$ term, then the equation describes a positive feedback or runaway process.

For a blending process (negative feedback or self-regulating process):

$$C_o{*}D_o{*}V = \int (C_i{*}D_i{*}F_i - C_o{*}D_o{*}F_o){*}dT \qquad \text{(C2)}$$

Divide through by $(F_o{*}D_o)$:

$$\frac{C_o{*}D_o{*}V}{D_o{*}F_o} = \int \left[ \frac{C_i{*}D_i{*}F_i}{D_o{*}F_o} - C_o \right]{*}DT \qquad \text{(C3)}$$

$$K_p = \frac{D_i{*}F_i}{D_o{*}F_o} = 1 \qquad \text{(C4)}$$

($D_i{*}F_i = D_o{*}F_o$ for gravity overflow or good level control.)

$$TC = \frac{D_o{*}V}{D_o{*}F_o} = \frac{V}{F_o} \qquad \text{(C5)}$$

where

$C_o$ = weight fraction concentration of the output stream
$C_i$ = weight fraction concentration of the input stream
$D_o$ = density of the output stream
$D_i$ = density of the input stream
$F_o$ = volumetric flow of the output stream
$F_i$ = volumetric flow of the input stream
$V$ = volume of the tank

A continuous concentration control process is self-regulating since there is an output stream whereas a batch concentration control process is integrating since there is no output stream ($F_o = 0$).

For a level process (no feedback or integrating process):

$$D_o{*}A{*}H = \int (D_i{*}F_i - D_o{*}F_o){*}dT \tag{C6}$$

$$W_i = D_i{*}F_i \tag{C7}$$

$$W_o = D_o{*}F_o \tag{C8}$$

Substitute Equations C7 and C8 into Equation C9:

$$A{*}D_o{*}H = \int (W_i - W_o){*}dT \tag{C9}$$

$$K_i = \frac{K_p}{TC} = \frac{1}{(A{*}D_o)} \tag{C10}$$

where

$H$ = tank level
$A$ = tank cross-sectional area
$D_i$ = density of the inlet flow
$D_o$ = density of the outlet flow
$F_i$ = volumetric flow in
$F_o$ = volumetric flow out
$W_i$ = mass flow in
$W_o$ = mass flow out
$K_i$ = integrator gain

The above process has no feedback because the flow out of the tank, $F_o$, does not depend on level $H$, assuming there is a pump on the discharge of the tank. If $F_i$ was used as the input instead of $W_i$, then the open loop error would be volumetric instead of mass flow and the integrator gain, $K_i$, would be equal to $1/A$. (The open loop error units must be consistent with the selected input variable units.)

For a low-range pressure process (negligible feedback or pseudointegrator), where the volumetric flow out is fixed by an induced draft fan:

$$D_o{*}V = \int (D_i{*}F_i - D_o{*}F_o){*}dT \tag{C11}$$

$$P_o = C_f{*}D_o{*}R{*}T \tag{C12}$$

(Equation C12 is the ideal gas law where $C_f$ is a unit conversion factor.)

Substitute Equation C12 into Equation C11:

$$\frac{(P_o{*}V)}{F_o} = \int \left[ \frac{C_f{*}R{*}T{*}D_i}{F_o}{*}F_i - P_o \right]{*}dT \tag{C13}$$

$$K_i = \frac{K_g}{TC} = \frac{C_f{*}R{*}T{*}D_i}{V} \tag{C14}$$

where

$P_o$ = low-range pressure to be controlled
$F_o$ = volumetric flow out
$F_i$ = volumetric flow in
$K$ = conversion factor to low-range pressure
$R$ = universal gas law constant
$T$ = absolute gas temperature
$V$ = gas volume

The low-range pressure process is a pseudointegrator because the time constant is large for large volumes. However the integrator gain is not small because $K_p$ is also large, which increases the open loop error and hence the closed loop errors. Notice that the input variable (disturbance), and therefore the open loop error, is volumetric flow, $F_i$. If the disturbance to be studied is gas density, $D_i$, then $F_i$ should be substituted for $D_i$ in the $K_p$ term in the equation for $K_i$.

For a low-range pressure process (negligible feedback or pseudointegrator) with the inlet and outlet volumetric flows determined by the pressure drop across flow resistances:

$$C^*P2 = \int\left[\frac{(P1 - P2)}{R1} + \frac{(P2 - P3)}{R2}\right]^*dT \quad \text{(C15)}$$

Collect terms over common denominator of $(R1 + R2)$:

$$TC^*P2 = \int(K1^*P1 + K3^*P3 - P2)^*dT \quad \text{(C16)}$$

$$TC = \left[\frac{R1^*R2}{R1 + R2}\right]^*C \quad \text{(C17)}$$

$$K1 = \frac{R2}{R1 + R2} \quad \text{(C18)}$$

$$K2 = \frac{R1}{R1 + R2} \quad \text{(C19)}$$

where

$P1$ = inlet pressure
$P2$ = pressure to be controlled
$P3$ = outlet pressure
$C$ = capacitance (see Appendix G)
$R1$ = inlet flow resistance (see Appendix G)
$R2$ = outlet flow resistance (see Appendix G)
$TC$ = time constant
$K1$ = process gain for an inlet pressure disturbance
$K2$ = process gain for an outlet pressure disturbance

Equations C16–C19 show that the time constant can become very large for either a large capacitance (volume) or for large inlet and outlet resistances (pressure drops). The integrator gain depends on the location of the disturbance and is inversely proportional to the flow resistance at that location and the capacitance of the volume. Thus the integrator gain is typically less than for the previous case where the outlet flow was fixed by an induced draft fan.

APPENDIX 

# Equation for Direct Synthesis Controller

The Laplace transfer functions for the loop are

$$\frac{O(s)}{E(s)} = \left[\frac{C(s)}{1 - C(s)}\right]*\left[\frac{K}{G(s)}\right] \quad \text{(D-1)}$$

$$C(s) = \left[\frac{1 - e^{-T/s}}{s}\right]*\left[\frac{e^{-TD_c/s}}{TC_c*s - 1}\right] \quad \text{(D-2)}$$

$$G(s) = \left[\frac{1 - e^{-T/s}}{s}\right]*\left[\frac{e^{-TD_o/s}}{TC_o*s - 1}\right] \quad \text{(D-3)}$$

where

$O(s)$ = Laplace transform of controller output
$E(s)$ = Laplace transform of controller error

$C(s)$ = Laplace transform of closed loop response with zero-order hold
$G(s)$ = Laplace transform of open loop response with zero-order hold
$T$ = sample period
$TD_c$ = closed loop response dead time
$TC_c$ = closed loop response time constant
$TD_o$ = open loop response dead time
$TC_o$ = open loop response time constant
$K$ = inverse of open-loop steady-state gain

Convert the closed and open loop response from Laplace to $Z$ transform:

$$C(s) = \left(1 - e^{-T/s}\right)*\left(\frac{1}{s} - \frac{1}{s + 1/TC_c}\right)*e^{-TD_c/s}*\frac{1}{TC_c} \quad \text{(D4)}$$

$$C(z) = \frac{Z-1}{Z}*\left(\frac{Z}{Z-1} - \frac{Z}{Z-A}\right)*Z^{-N}*\frac{1}{TC_c} \quad \text{(D5)}$$

$$C(z) = \frac{(1-A)*Z^{-(N+1)}}{(1 - A*Z^{-1})*TC_c} \quad \text{(D6)}$$

$$A = e^{-T/TC_c} \quad \text{(D7)}$$

$$G(s) = \left(1 - e^{-T/s}\right)*\left(\frac{1}{s} - \frac{1}{s + 1/TC_o}\right)*e^{-TD_o/s}*\frac{1}{TC_o} \quad \text{(D8)}$$

$$G(z) = \frac{Z-1}{Z}*\left(\frac{Z}{Z-1} - \frac{Z}{Z-B}\right)*Z^{-N}*\frac{1}{TC_o} \quad \text{(D9)}$$

$$G(z) = \frac{(1-B)*Z^{-(N+1)}}{(1 - B*Z^{-1})*TC_o} \quad \text{(D10)}$$

$$B = e^{-T/TC_o} \quad \text{(D11)}$$

Substitute Equations D-6 and D-10 into D-1:

$$\frac{O(z)}{E(z)} = \frac{K^*(1 - B)^*(1 - A^*Z^{-1})}{(1 - A)^*(1 - B^*Z^{-1} - (1 - B)^*Z^{-(N+1)})} * \frac{1}{TC_c^*TC_o} \quad \text{(D12)}$$

Replace all constant terms with a single constant:

$$\frac{O(z)}{E(z)} \quad \frac{K'^*(1 - A^*Z^{-1})}{(1 - B^*Z^{-1} - (1 - B)^*Z^{-(N+1)})} \quad \text{(D13)}$$

Multiply and convert to the finite difference equation by using the negative power of $Z$ as the prior sample number:

$$O(z)^*(1 - B^*Z^{-1} - (1 - B)^*Z^{-(N+1)}) = E(z)^*K'^*(1 - A^*Z^{-1}) \quad \text{(D14)}$$

$$O_n - B^*O_{n-1} - (1 - B)^*O_{n-N-1} = K'^*(E_n - A^*E_{n-1}) \quad \text{(D15)}$$

$$O_n = B^*O_{n-1} + (1 - B)^*O_{n-N-1} + K'^*(E_n - A^*E_{n-1}) \quad \text{(D16)}$$

where

$n$ = subscript denoting value from present sample period
$n - 1$ = subscript denoting value from one sample period ago
$n - N$ = subscript denoting value from $N$ sample periods ago

Equation D16 is identical to Equation 11.6.

APPENDIX E

# *Equations for Liquid Composition Control Dynamics*

The dead time and time constant of tanks, static mixers, and pipelines for liquid composition control can be estimated by Equations E1 and E2.

$$TD = \frac{V}{F + F_a} \tag{E1}$$

$$TC = \frac{V}{F} - TD \tag{E2}$$

$$\frac{TD}{TC} = \frac{F}{F_a} \tag{E3}$$

$$\frac{TD}{TD + TC} = \frac{F}{F + F_a} \tag{E4}$$

where

$TD$ = equipment dead time for liquid composition control
$TC$ = equipment time constant for liquid composition control
$V$ = liquid volume of the equipment
$F$ = throughput liquid flow rate
$V/F$ = residence time
$F_a$ = equivalent flow due to agitation

For axial impeller agitators:

$$F_a = 7.5^*N_q^*N_s^*D_t^3 \tag{E5}$$

$$N_q = \frac{0.4}{\left[\frac{D_i}{D_t}\right]^{0.55}} \tag{E6}$$

where

$F_a$ = equivalent flow due to agitation (rpm)
$N_q$ = axial impeller discharge coefficient
$N_s$ = axial impeller speed (rpm)
$D_i$ = axial impeller diameter (feet)
$D_t$ = tank diameter (feet)

For self-regulating processes, the peak error is proportional to the dead-time to time-constant ratio (Equation E3) for small dead times and is proportional to the dead-time to residence-time ratio (Equation E4) for large dead times (see Equations B5 and B6). The ratio of throughput flow to agitation flow $F/F_a$ is typically 0.05 for a vertical well-mixed tank with an axial impeller, 1.5 for laminar segregated flow in a pipeline, and 3.0 for a static pipeline mixer. The ratio $F/F_a$ for turbulent flow in a pipeline is slightly less than that for the static mixer. The ratio $F/F_a$ becomes smaller as the velocity profile in a pipeline becomes more parabolic. The fact that the static mixer approaches plug flow is undesirable from a control response viewpoint but may be desirable for the use of the mixer as a plug flow reactor. Each infinitesimal reactant volume will have the same residence time (Bor, 1971).

Equations E1 and E2 lead to some interesting conclusions about the use of well-mixed tanks for composition control. If the tank volume is increased but the flows $F$ and $F_a$ are kept constant, the peak error is the same but the accumulated error is increased in proportion to the increase in tank volume (the peak error will also increase if the disturbance time constant is not negligible; see Chapter 8). If the tank volume is made exceptionally large, the dead time may be so large that the minimum reset setting (maximum integral time) may cause a reset cycle. If the agitation flow were increased in proportion to the tank volume, the peak error would be reduced and the accumulated error would be about the same. (The agitation flow is much larger than the throughput flow in Equation E1 for a well-mixed tank.) If the increased tank volume does not have a control loop on it, the tank time constant will slow down step disturbances and filter oscillatory disturbances from control loops upstream and for control loops downstream, provided there is some turbulence in the tank. Thus exceptionally large tank volumes improve composition control loop performance if the loops are upstream or downstream, but not on the tank, unless the agitation flow is scaled up with the tank volume. If the tank is upstream of the loop, the reagent or coolant usage will be reduced because of filtering of disturbance oscillations.

APPENDIX F

# Equations for Equivalent Noninteractive Time Constants

The interactive time constants must be broken up into their capacitance and resistance components in order to calculate the equivalent noninteractive time constants. The quadratic equation can then be used to find the two equivalent noninteractive time constants for two capacitances that have an inlet and outlet resistance and are separated by a resistance (Harriot, 1964).

$$TC1 = \frac{2*A}{B - [B^2 - 4*A*C]^{0.5}} \tag{F1}$$

$$TC2 = \frac{2*A}{B + [B^2 - 4*A*C]^{0.5}} \tag{F2}$$

$$A = R1*C1*R2*C2*R3 \tag{F3}$$

$$B = R1*C1*R3 + R1*C1*R2 + R2*C2*R3 + R1*C2*R3 \tag{F4}$$

$$C = R1 + R2 + R3 \tag{F5}$$

where

$TC1$ = first equivalent noninteractive time constant
$TC2$ = second equivalent noninteractive time constant
$A$ = first term of the quadratic equation
$B$ = second term of the quadratic equation
$C$ = third term of the quadratic equation
$R1$ = inlet resistance
$R2$ = intermediate resistance
$R3$ = outlet resistance
$C1$ = first capacitance
$C2$ = second capacitance

The capacitances and resistances for a gas pressure system are defined in Appendix G. The products $R1*C1$ and $R2*C2$ are not the time constants of the individual tanks or sections of pipeline in Appendix G because each volume has two parallel flow resistances associated with it so that the interactive time constants are actually

$$\frac{C1}{\left[\frac{1}{R1}+\frac{1}{R2}\right]}$$

and

$$\frac{C2}{\left[\frac{1}{R2}+\frac{1}{R3}\right]}$$

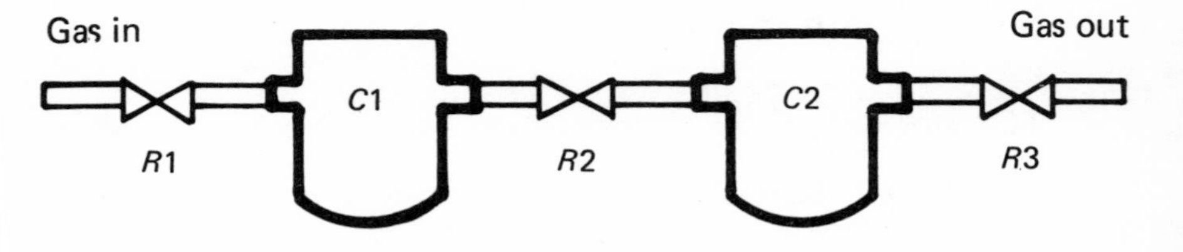


FIGURE F1
Two interactive time constants.

APPENDIX G

# Equations for Gas Pressure Control Time Constants

The interactive time constants of multiple tanks or sections of pipelines (volumes) in series separated by flow resistances can be estimated by Equations G1–G4 for turbulent and sonic flow (Harriot, 1964).

$$TC = \frac{C}{\sum\left[\frac{1}{R_j}\right]} \tag{G1}$$

$$C = \frac{V}{P_a} \tag{G2}$$

For turbulent flow:

$$R_j = \frac{2^*\Delta P_j}{F_j} \tag{G3}$$

For sonic flow:

$$R_j = \frac{P_j}{F_j} \tag{G4}$$

where

$TC$ = interactive time constant for the tank or section, minutes
$C$ = capacitance of the volume, scf/psi
$R_j$ = $j$th flow resistance connected to the volume, psi/scfm
$V$ = volume of the tank or section of pipeline, scf
$P_a$ = atmospheric pressure, psi ($P_a$ = 14.7)
$\Delta P_j$ = pressure drop across the $j$th flow resistance, psi
$P_j$ = inlet pressure at the $j$th flow resistance, psi
$F$ = flow across the resistance, scfm

If the intermediate flow resistance pressure drops for turbulent flow are less than 1 percent of the inlet and outlet flow resistance pressure drops, then all the individual tank or section volumes can be summed and substituted in Equation G2 to give a single time constant. The universal gas sizing equation can be used to estimate the $C_v$ of the fitting or the valve necessary to prevent the creation of multiple time constants and the associated dead time.

$$C_v > \left[\frac{F}{P^*\left[\frac{(0.01^*\Delta P)}{P}\right]^{0.5}}\right]^*\left[\frac{(G^*T)}{(520)}\right]^{0.5} \tag{G5}$$

where

$C_v$ = flow coefficient (inverse of flow resistance)
$F$ = gas flow, scfm
$P$ = inlet pressure, psia
$\Delta P$ = pressure drop at the discharge of the last volume, psi
$G$ = gas specific gravity
$T$ = absolute gas temperature, degrees Roentgen

APPENDIX 

# *FORTRAN Listing of Subroutine to Tune a PID Controller*

```
      SUBROUTINE ZTUNE(Y,P)
      DIMENSION Y(3),P(11)
C
C  Y = AN OUTPUT ARRAY WHOSE ELEMENTS ARE:
C    Y(1) = PROPORTIONAL BAND AND ACTION - POSITIVE FOR DIRECT ACTION
C           AND NEGATIVE FOR REVERSE ACTION (PER CENT)
C    Y(2) = RESET (INTEGRAL) SETTING - (REPEATS/PROGRAM TIME UNIT)
C    Y(3) = RATE (DERIVATIVE) SETTING - (PROGRAM TIME UNITS)
C
C  P = A PARAMETER ARRAY WHOSE ELEMENTS ARE:
C    P(1) = CONTROLLER TYPE INDICATOR SET AS FOLLOWS:
C         0. = PROPORTIONAL ONLY (P)
C         1. = PROPORTIONAL-RESET (PI)
C         2. = PROPORTIONAL-RESET-RATE (PID)
C         3. = PROPORTIONAL-RATE (PD)
C    P(2) = PROCESS TYPE INDICATOR SET AS FOLLOWS:
C         0. = SELF-REGULATING
C         1. = INTEGRATING
C         2. = RUNAWAY
```

```
C     P(3) = PRODUCT OF VALVE, PROCESS, AND TRANSMITTER GAIN
C            (% CHANGE IN MEASUREMENT DIVIDED BY % CHANGE IN
C            VALVE POSITION) CONTROLLER ACTION VIA Y(1) WILL
C            HAVE THE OPPOSITE SIGN
C     P(4) = VALVE TIME CONSTANT - (PROGRAM TIME UNITS)
C     P(5) = VALVE DEADTIME - (PROGRAM TIME UNITS)
C     P(6) = MEASUREMENT TIME CONSTANT - (PROGRAM TIME UNITS)
C     P(7) = MEASUREMENT DEADTIME - INCLUDE 1.5*TS FOR A CHROMATOGRAPH
C            AND 0.7*TS FOR A DIGITAL CONTROLLER WHERE TS IS THE
C            SAMPLE TIME - (PROGRAM TIME UNITS)
C     P(8) = THE LARGEST SELF-REGULATING PROCESS TIME CONSTANT-
C            - (PROGRAM TIME UNITS)
C     P(9) = PROCESS DEADTIME (MUST BE USED) - P(9) IS THE SUMMATION
C            OF ANY TIME CONSTANTS SMALLER THAN P(10) PLUS
C            TRANSPORTATION DELAYS OR CAN BE DETERMINED GRAPHICALLY
C            - (PROGRAM TIME UNITS)
C    P(10) = NEXT LARGEST SELF-REGULATING PROCESS TIME CONSTANT
C            P(10) IS DETERMINED GRAPHICALLY THE SAME AS P(8)
C            BUT FOR AN INTERMEDIATE OPEN LOOP RESPONSE
C            - (PROGRAM TIME UNITS)
C    P(11) = RUNAWAY PROCESS TIME CONSTANT - MUST BE LARGER THAN
C            THE LARGEST SELF-REGULATING TIME CONSTANT AND TOTAL
C            LOOP DEADTIME TO PREVENT INSTABILITY IF A RUNAWAY
C            PROCESS IS SPECIFIED (P(2)=2.) - (PROGRAM TIME UNITS)
C            IF P(2) = 2.:
C            P(11) > P(8)
C            P(11) > P(4) + P(5) + P(6) + P(7) + P(9) + P(10)
C
C      MODE |    P     |   PI*    |   PID    |   PD     |
C      -----|----------|----------|----------|----------|
C      Y(1) | 1.8*PBU  | 2.0*PBU  | 1.5*PBU  | 1.4*PBU  |
C      Y(2) |    0.    | 1.2/TU   | 2./TU    |   0.     |
C      Y(3) |    0.    |   0.     | .125*TU  | .075*TU  |
C
C     * PI MODE IS NOT RECOMMENDED FOR NON-SELF-REGULATING
C       PROCESSES WITH SLOW CONTROL VALVES OR MEASUREMENTS.
```

```
C     CALCULATE THE FIRST ORDER PLUS DEADTIME APPROXIMATION
C
      TD = P(5) + P(7) + P(9)
      TCMAX = AMAX1(P(4),P(6),P(8),P(10))
      TCMAX = TCMAX*(0.9999)
      TC = TCMAX
      FO = 1.0
      IF (P(2).GT.0.0) FO = 1.2
      DO  5 I = 4,10,2
      TCMIN = P(I)
      X = TCMIN/TCMAX
      IF (X.EQ.0.0) X = 0.0001
      IF (X.EQ.1.0) X = 0.9999
      IF (X.GT.1.0) GO TO 5
      Z = 1./(1. - X)
      F = FO + 1./X - 1./X**Z - ALOG(X)/(1. - X)
      TD = TD + F*TCMIN
      TC = TC + (1. - F)*TCMIN
    5 CONTINUE
      J = P(2) + 1
      GO TO (10,20,30), J
C
C     CALCULATE THE ULTIMATE PERIOD AND PB FOR SELF-REGULATING PROCESS
C
   10 TU = 2.*(1. + (TC/(TC+TD))**0.65)*TD
      PBU = -P(3)*100./(((6.28*TC/TU)**2.+1.)**0.5)
      GO TO 40
C
C     CALCULATE THE ULTIMATE PERIOD AND PB FOR AN INTEGRATING PROCESS
C
   20 TU = 4.*(1. + (TC/TD)**0.65)*TD
      PBU = -P(3)*TU*100./(6.28*((6.28*TC/TU)**2.+1.)**0.5)
      GO TO 40
C
```

```
C     CALCULATE THE ULTIMATE PERIOD AND PB FOR A RUNAWAY PROCESS
C
   30 TU = 4.*(1. + (((P(11)+TC)*P(11)*TC)/((P(11)-TC)*(P(11)
     $      -TD)*TD))**0.65)*TD
      PBU = -P(3)*100./((((6.28*TC/TU)**2.+1.)**0.5)*
     $       (((6.28*P(11)/TU)**2.+1.)**0.5))
C
C     CALCULATE CONTROLLER MODE SETTINGS DEPENDING ON CONTROLLER TYPE
C
   40 IF (P(1).EQ.0.0) Y(1) = 1.8*PBU
      IF (P(1).EQ.1.0) Y(1) = 2.0*PBU
      IF (P(1).EQ.2.0) Y(1) = 1.5*PBU
      IF (P(1).EQ.3.0) Y(1) = 1.4*PBU
      Y(2) = 0.0
      Y(3) = 0.0
      IF (P(1).EQ.1.0) Y(2) = 1.2/TU
      IF (P(1).EQ.2.0) Y(2) = 2.0/TU
      IF (P(1).EQ.2.0) Y(3) = 0.125*TU
      IF (P(1).EQ.3.0) Y(3) = 0.075*TU
      RETURN
      END
```

APPENDIX I

# ACSL Listing of Dynamic Simulation Program of Cascade Control

```
PROGRAM CASCADE CONTROL PERFORMANCE TEST
'"R" DESIGNATES REMOTE SP SECONDARY AND "L" LOCAL SP PRIMARY LOOP'
'"XXX=0." DESIGNATES SINGLE OVERALL LOOP AND "XXX=1." CASCADE LOOP'
'"RRR1=0." DESIGNATES AN INTEGRATING SECONDARY (REMOTE SP) LOOP'
'"LLL1=0." DESIGNATES AN INTEGRATING PRIMARY (LOCAL SP) LOOP'
'"RRR1=1." DESIGNATES A RUNAWAY SECONDARY (REMOTE SP) LOOP'
'"LLL1=1." DESIGNATES A RUNAWAY PRIMARY (LOCAL SP) LOOP'
'"RRR2=0." DESIGNATES A SELF-REGULATING SECONDARY (REMOTE SP) LOOP'
'"LLL2=0." DESIGNATES A SELF-REGULATING PRIMARY (LOCAL SP) LOOP'
'PROGRAM ASSUMES PID CONTROLLER AND NEGLIGABLE VALVE DYNAMICS'
'THE MODE SETTINGS FROM THE TUNE MACRO CAN BE ALTERED BY FACTORS'
'THE TUNE MACRO USES THE ZTUNE SUBROUTINE DOCUMENTED IN APPENDIX H'
'THE CNTRL MACRO USES A ZCNTRL SUBROUTINE DEVELOPED BY MONSANTO'
'THE CNTRL MACRO SIMULATES AN INDUSTRIAL THREE MODE CONTROLLER'
```

```
INITIAL
ARRAY PR(11),PL(11),KR(9),KL(9),YR(3),YL(3)
MACRO TUNE(Y,P)
PROCEDURAL(Y=P)
CALL ZTUNE(Y,P)
END
MACRO END
CONSTANT KR=-100.,0.,0.,1.,1.,0.,1.,0.,1.
CONSTANT KL=-100.,0.,0.,1.,1.,0.,1.,0.,1.
CONSTANT PR1IC=0.5,PR2IC=0.5,PR3IC=0.0,MRIC=0.5,MRSP=0.5
CONSTANT PL1IC=0.5,PL2IC=0.5,PL3IC=0.0,MLIC=0.5,MLSP=0.5
CONSTANT PFR=1.,IFR=1.,DFR=1.,RSD=0.2
CONSTANT PFL=1.,IFL=1.,DFL=1.,LSD=0.0
CONSTANT KIC=0.5,KKK=2.2,XXX=0.0
CONSTANT TCF=1.0,TDF=1.0,FFR=0.,FFL=0.
PBO=ABS(KL(1))
EX1=0.
EX2=0.
FFF=0.
PR(1)=2.0
PR(2)=FFR
PR(3)=1.0
PR(4)=0.0
PR(5)=0.0
PR(6)=0.0
PR(7)=0.0
PR(8)=1.0*TCF
PR(9)=0.0
PR(10)=0.25*TDF
PR(11)=5.0
PL(1)=2.0
PL(2)=FFL
PL(3)=1.0
PL(4)=0.0
PL(5)=0.0
```

```
PL(6)=0.0
PL(7)=0.0
PL(8)=1.0
PL(9)=0.0
PL(10)=0.25
PL(11)=5.0
RRR1=PR(2) - 1.
GR=PR(3)
TDPR=PR(10)
TCPR1=PR(9)
TCPR2=PR(8)
TCPR3=PR(11)
IF (RRR1.LT.+0.1) TCPR3=1.
RRR2=0.
IF (RRR1.GT.-0.1) RRR2=1.
TUNE(YR,PR)
KR(1)=PFR*YR(1)
KR(2)=IFR*YR(2)
KR(3)=DFR*YR(3)
KR(4)=XXX
LLL1=PL(2) - 1.
GL=PL(3)
TDPL=PL(10)
TCPL1=PL(9)
TCPL2=PL(8)
TCPL3=PL(11)
IF (LLL1.LT.+0.1) TCPL3=1.
LLL2=0.
IF (LLL1.GT.-0.1) LLL2=1.
PL(2)=FFR*(1.0-XXX)+FFL
PL(9)=TCPL1+(1.-XXX)*TCPR2
PL(10)=TDPL+(1.-XXX)*TDPR+0.5*0.9*XXX/YR(2)
AAA=0.4
BBB=0.4
CCC=0.2
```

```
IF (FFR.EQ.1.0) AAA=0.0
IF (FFR.EQ.1.0) BBB=1.0
IF (FFR.EQ.1.0) CCC=-1.0
IF (FFR.EQ.2.0) AAA=0.0
IF (FFR.EQ.2.0) BBB=1.0
IF (FFR.EQ.2.0) CCC=-1.0
IF (FFL.EQ.1.0) AAA=0.0
IF (FFL.EQ.1.0) BBB=0.0
IF (FFL.EQ.1.0) CCC=0.0
IF (FFL.EQ.2.0) AAA=0.0
IF (FFL.EQ.2.0) BBB=0.0
IF (FFL.EQ.2.0) CCC=0.0
PFL=1.0+(AAA*TDF+BBB*TCF-CCC)*XXX
TUNE(YL,PL)
KL(1)=PFL*YL(1)
KL(2)=IFL*YL(2)
KL(3)=DFL*YL(3)
CONSTANT TSD=0.1
CONSTANT TSTOP=20.
ALGORITHM IALG=1
CINTERVAL CINT=0.1
END
DYNAMIC
DERIVATIVE
CL=CNTRL(KIC,KL,ML,MLSP)
CR=CNTRL(KIC,KR,MR,MRSP)
MRSP=CL
C=XXX*CR+(1.-XXX)*CL
PR1=DELAY(C,PR1IC,TDPR,5000)+RSD*STEP(TSD)
PR2=REALPL(TCPR2,GR*PR1,PR2IC)
PR3=INTEG(RRR2*(PR2-MLSP+RRR1*PR3)/TCPR3,PR3IC)
MR=RRR2*(PR3+MLSP)+(1.-RRR2)*PR2
O=MR
PL1=DELAY(O,PL1IC,TDPL,5000)+LSD*STEP(TSD)
PL2=REALPL(TCPL2,GL*PL1,PL2IC)
```

```
PL3=INTEG(LLL2*(PL2-MLSP+LLL1*PL3)/TCPL3,PL3IC)

ML=LLL2*(PL3+MLSP)+(1.-LLL2)*PL2

EI1=INTEG(ML-MLSP,0.)

EI2=INTEG(ABS(ML-MLSP),0.)

PBL=ABS(YL(1))

PBR=ABS(YR(1))

AFL=1./((3.14*PL(8)*YR(2))**2.+1.)**0.5

CX1=(KKK*(PBR/100.)*AFL*XXX+(PBL/100.)*(1.-XXX))*(RSD+LSD)

CX2=(KKK*(PBR/100.)*(PBL/100.)*AFL*XXX+(PBL/100.)*(1.-XXX))*(RSD+LSD)

CI1=CX1/YL(2)

CI2=CX2/YL(2)

PROCEDURAL (EX1,EX2=ML,MLSP,FFF)

IF (ML-MLSP.GT.EX1) EX1=ML-MLSP

IF (ML-MLSP.LT.-0.0001) FFF=1.

IF (FFF.GT.0.0.AND.ML-MLSP.GT.EX2) EX2=ML-MLSP

END

END

TERMT(T.GE.TSTOP)

END

TERMINAL

END

END

SET PRN=9

SET TTLCPL=.T.

SET TITLE = 'SINGLE TCF=1., TDF=1., FFR=0., & FFL=0.'

SET NGXPPL=10,NGYPPL=10,NPXPPL=50,NPYPPL=50

PREPAR T,C,MR,ML

SET XXX=0.,TCF=1.,TDF=1.,FFR=0.,FFL=0.,TSTOP=20.

PROCED RUN

START

PLTDIN

SET XINCPL=4., YINCPL=4.

PLOT 'XHI'=TSTOP,MR,ML

DISPLY XXX,KKK
```

```
DISPLY AAA,BBB,CCC
DISPLY C,MR,ML
DISPLY FFR,FFL
DISPLY TCF,TDF
DISPLY EI1,EI2
DISPLY CI1,CI2
DISPLY EX1,EX2
DISPLY CX1,CX2
DISPLY PR,PL
DISPLY YR,YL
DISPLY KR,KL
END
RUN
SET TITLE = 'CASCADE TCF=1., TDF=1., FFR=0., & FFL=0.'
SET XXX=1.,TCF=1.,TDF=1.,TSTOP=10.
RUN
SET TITLE = 'SINGLE TCF=.8, TDF=1., FFR=0., & FFL=0.'
SET XXX=0.,TCF=.8,TDF=1.,TSTOP=20.
RUN
SET TITLE = 'CASCADE TCF=.8, TDF=1., FFR=0., & FFL=0.'
SET XXX=1.,TCF=.8,TDF=1.,TSTOP=10.
RUN
SET TITLE = 'SINGLE TCF=.6, TDF=1., FFR=0., & FFL=0.'
SET XXX=0.,TCF=.6,TDF=1.,TSTOP=20.
RUN
SET TITLE = 'CASCADE TCF=.6, TDF=1., FFR=0., & FFL=0.'
SET XXX=1.,TCF=.6,TDF=1.,TSTOP=10.
RUN
SET TITLE = 'SINGLE TCF=.4, TDF=1., FFR=0., & FFL=0.'
SET XXX=0.,TCF=.4,TDF=1.,TSTOP=20.
RUN
SET TITLE = 'CASCADE TCF=.4, TDF=1., FFR=0., & FFL=0.'
SET XXX=1.,TCF=.4,TDF=1.,TSTOP=10.
RUN
SET TITLE = 'SINGLE TCF=.2, TDF=1., FFR=0., & FFL=0.'
SET XXX=0.,TCF=.2,TDF=1.,TSTOP=20.
```

```
RUN
SET TITLE = 'CASCADE TCF=.2, TDF=1., FFR=0., & FFL=0.'
SET XXX=1.,TCF=.2,TDF=1.,TSTOP=10.
RUN
SET TITLE = 'SINGLE TCF=1., TDF=.6, FFR=0., & FFL=0.'
SET XXX=0.,TCF=1.,TDF=.6,TSTOP=20.
RUN
SET TITLE = 'CASCADE TCF=1., TDF=.6, FFR=0., & FFL=0.'
SET XXX=1.,TCF=1.,TDF=.6,TSTOP=10.
RUN
SET TITLE = 'SINGLE TCF=.8, TDF=.6, FFR=0., & FFL=0.'
SET XXX=0.,TCF=.8,TDF=.6,TSTOP=20.
RUN
SET TITLE = 'CASCADE TCF=.8, TDF=.6, FFR=0., & FFL=0.'
SET XXX=1.,TCF=.8,TDF=.6,TSTOP=10.
RUN
SET TITLE = 'SINGLE TCF=.6, TDF=.6, FFR=0., & FFL=0.'
SET XXX=0.,TCF=.6,TDF=.6,TSTOP=20.
RUN
SET TITLE = 'CASCADE TCF=.6, TDF=.6, FFR=0., & FFL=0.'
SET XXX=1.,TCF=.6,TDF=.6,TSTOP=10.
RUN
SET TITLE = 'SINGLE TCF=.4, TDF=.6, FFR=0., & FFL=0.'
SET XXX=0.,TCF=.4,TDF=.6,TSTOP=20.
RUN
SET TITLE = 'CASCADE TCF=.4, TDF=.6, FFR=0., & FFL=0.'
SET XXX=1.,TCF=.4,TDF=.6,TSTOP=10.
RUN
SET TITLE = 'SINGLE TCF=.2, TDF=.6, FFR=0., & FFL=0.'
SET XXX=0.,TCF=.2,TDF=.6,TSTOP=20.
RUN
SET TITLE = 'CASCADE TCF=.2, TDF=.6, FFR=0., & FFL=0.'
SET XXX=1.,TCF=.2,TDF=.6,TSTOP=10.
RUN
SET CMD=5
```

APPENDIX 

# FORTRAN Listing of Subroutine to Estimate the Effective Dead Time and Time Constant for Equal Noninteractive Time Constants

```
      SUBROUTINE ZEQUIV(Z,F)
      DIMENSION Z(3),F(2),A(5)
      REAL N
      DATA A/-0.06695235,0.2595017,0.08436275,-0.04161446,0.004497025/
C
C     CALCULATED OUTPUTS:
C     Z(1) IS THE EFFECTIVE TIME CONSTANT (P(8) IN ZTUNE SUBROUTINE *)
C     Z(2) IS THE EFFECTIVE DEADTIME (P(9) IN ZTUNE SUBROUTINE *)
C     Z(3) IS THE EFFECTIVE DEADTIME FACTOR
C     * - ZTUNE SUBROUTINE IS DOCUMENTED IN APPENDIX H
C     PARAMETER INPUTS:
C     F(1) IS THE INDIVIDUAL TIME CONSTANT VALUE
C     F(2) IS THE NUMBER OF EQUAL TIME CONSTANTS (REAL NUMBER)
C
      TC = F(1)
      N = F(2)
      Y = A(1) + A(2)*ALOG(N) + A(3)*ALOG(N)**2 + A(4)*ALOG(N)**3
     $    + A(5)*ALOG(N)**4
      Z(1) = ((1.-Y)*N*TC)/(1.+Y)
      Z(2) = Y*N*TC
      Z(3) = Y
      RETURN
      END
```

# Index


Accumulated error
definition 24
equation 26
Adaptive control (see self-tuning)
Agitated vessel
deadtime 36, 237
time constant 36, 237
Aliasing 21, 92
Analog controller 88
Analyzer deadtime 114
Anti-reset windup 93

Batch controller 200
Biological reactor
control loop example 79, 129
runaway time constant 37
Boiler drum level
control loop example 69, 123
feed forward control 197
Boiler feedwater
control loop example 66, 121
feed forward control 197

Cascade control 183
Characterization
mode 199
signal 198
Compressor
speed runaway response 55
surge control valve 139
surge disturbance 158
surge measurement 107
Conductivity nonlinearity 169
Combustion control
carbon monoxide control 167, 206
furnace pressure interaction 180
opacity control 168
Concentration control
biological reactor example 79, 129
deadtime 36, 237
time constant 36, 237
Controllers
analog 88
definition of modes 3
digital 90
direct synthesis controller 202, 233
dynamic matrix controller 211
parallel 83
self-tuning controller 207
series 85
typical mode settings 13
typical tuning methods 14

Deadband (see hysteresis)
Deadtime
compensator 204
definition 9
from agitated vessel 36, 237
from chromatograph 114
from control valve 141
from digital controller 90
from equal time constants 39
from pneumatic tubing 111
from sample transportation 112
from small time constants 38
from static mixer 36, 237
from thermowell 109
graphical method 33

Decoupling 179
Derivative mode 6, 44
Detuning
cascade 194
interaction 182
noise 160
positioner 139
Digital controller 90
Digital control valve 147
Direct synthesis controller 202, 233
Dynamic matrix control 211

Error
accumulated 24
peak 30
Error squared controller 199
Exothermic reactor
runaway time constant 36
pilot plant scale up 57
control loop example 76, 126

Feedforward control 195
First order plus deadtime method 16
Furnace pressure loop example 71, 123, 154, 161

Gain (steady state)
conductivity 169
definition 9
installed valve characteristic 166
integrator 9, 50
heat exchanger 168
matrix 179
open loop 17, 29, 165
pH 169

Hysteresis 142, 149

Ideal controller 83
Integral mode, 4, 44
Integrated error
absolute 24
algebraic 24
squared 24
Integrating processes 34, 47
Inverse response 14, 172, 195

Laminar flow control valve 147
Lead-lag 172, 195
Limit cycle 143

Model predictive control 210
Modes
characterization 199
definition 3
interaction 112
typical settings 13

Negative feedback time constant 9
Noise
detuning 160
interaction 182
stochastic 208
Non-self-regulating processes
graphical method 34
integrating 47
runaway 53

Parallel controller 83
Peak error 30
Period
loop for slow measurement 102
loop for slow valve 137
ultimate 4, 15, 43
pH loop example 61, 117, 173
pH titration curve 9, 169, 199
Positive feedback time constant 9
Pressure control
time constant 36

furnace example 71, 123, 154, 161
Proportional band
definition 3
window 50, 56, 60
Proportional mode 3
Pulse interval control 149

Rangeability 145
Rate mode 6, 44
Rate limited exponential 133
Real controller 85
Reset mode 4, 44
Runaway processes 34, 53
Runaway time constant
exothermic reactor 36
biological reactor 37

Sample time
chromatograph 114
digital controller 90
Self-regulating processes 34, 42
Self-Tuning controller 207
Series controller 85
Signal characterization 198
Smith Predictor 204
Surge
Controller reset action 98, 158
Control valve 139
Disturbance speed 158
Measurement 107

Time constant
definition 9
for agitated vessel 36, 237
for biological reactor 37
for control valve 133
for disturbance 157
for exothermic reactor 36
for feedforward signal 196
for furnace 36, 243
for heat exchanger 36
for noise 160
for pH electrode 107
for pipeline 36, 237
for pneumatic tubing 112
for process heater 37
for static mixer 36, 237
for temperature sensor 110
for thermowell 109
for transmitter 111
graphical method 34
interactive 40, 241
Time delay (see deadtime)
Transportation delay (see deadtime)

Ultimate oscillation method 14
Ultimate period 4, 15, 43

Velocity limiting (see rate limiting)

Waste treatment (see pH loop example)
Window of proportional bands 50, 56, 60